Richard Hammond

All about PHYSICS

Senior editor Ben Morgan
Senior art editor Claire Patané
Category publisher Mary Ling
Art director Jane Bull
Publishing manager Sue Leonard
Managing art editor Clare Shedden
Editors Elizabeth Haldane, Zahavit Shalev, Susan Watt

REVISED EDITION
Senior editors Fleur Star, Bharti Bedi
Senior art editor Spencer Holbrook
Assistant editors Tejaswita Payal, Sheryl Sadana
Assistant art editor Nidhi Rastogi
DTP designer Sachin Gupta
Senior DTP designer Harish Aggarwal
Jacket designer Suhita Dharamjit
Jackets assistant Claire Gell
Jacket design development manager Sophia MTT
Producer, pre-production Gillian Reid
Producer Vivienne Yong
Managing editors Linda Esposito, Kingshuk Ghoshal
Managing art editors Philip Letsu, Govind Mittal
Publisher Andrew Macintyre
Publishing director Jonathan Metcalf
Associate publishing director Liz Wheeler
Design director Stuart Jackman

First published as Can You Feel the Force? in hardback in Great Britain in 2006
This edition first published in 2015 by
Dorling Kindersley Limited
80 Strand, London WC2R 0RL

10 9 8 7 6 5 4 3 2
002 – 285064 – 09/15

A CIP catalogue record for this book
is available from the British Library.

ISBN 978-0-2412-0655-3

Printed and bound in China

Discover more at
www.dk.com

It's no secret that I love cars – and bikes, and planes, and speedboats, and hovercraft and, well, anything that moves really. And the reason I love them? Because they're all about action – stuff happening. When I drive a car round a racetrack, I can feel things going on all around me: the tyres scrabbling for grip, the acceleration pushing me back in the seat, the seatbelt straining when I hit the brakes and fly forward. In other words, there are a whole lot of forces going on. And that's physics.

Physics is the action department of science. When a car crashes, an apple falls out of a tree, or lightning strikes, the laws of physics can tell you what's going on. Biology and chemistry might be able to tell you why an apple tastes the way it does, but only physics can explain what happens if you throw it at a brick wall at 200 mph.

Of course, it's not all fast cars and squashed apples. In fact, physics is about, well, *everything* – from the inconceivably tiny bits of stuff the Universe is made of to the inconceivable vastness of the Universe itself. Physics is also the weird end of science. It's about the invisible, the unexplained, and the downright strange. Think of the slippery force that pushes magnets apart when you try to press two north poles together. What *is* that force and why is it there?

You might think scientists have got everything worked out, but the truth is that science is full of mysteries and questions, and that's why this book is full of questions too. Most of them have easy answers, but some have no answer yet. Some might surprise you, others might shock, and some are there just to make you think... We hope you enjoy them.

RICHARD HAMMOND

CONTENTS

In the BEGINNING

Can you FEEL THE FORCE?

What's THE MATTER?

Can you SEE THE LIGHT?

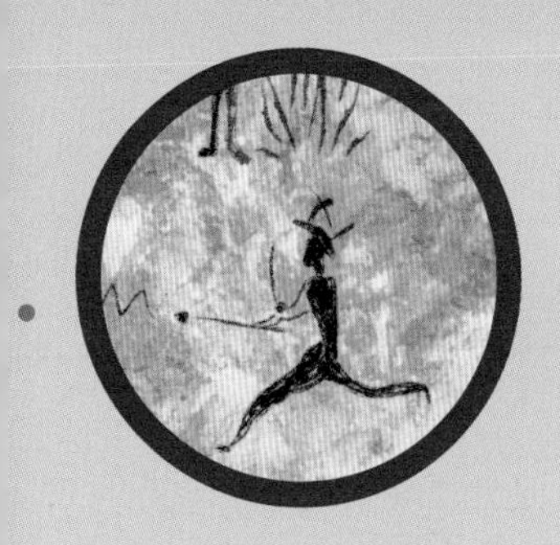

Greece is the word 8
The Dark Ages 12
Into the light 14
Galileo's world 16
Newton's Universe 18

What is a force? 22
It's the law! 24
Can you do physics on a bike? 26
What causes friction? 28
How fast can you go? 30
G force 32
Energy 34
How can you make forces bigger? .. 36
How do bicycles work? 38
How fast can you fall? 40
How do planes stay in the air? 42
Why do golf balls have dimples? ... 44
What's the best shape for a car? 46
Why do balls bounce? 48
Can you lie on a bed of nails? 50

What is matter made of? 54
What's inside an atom? 56
Why do balloons stick to the wall?.. 58
Shocking experiments 60
How do magnets work? 62
Can you feel the heat? 64
States of matter 66
What shape is a raindrop? 68
Can you walk on custard? 70
How does a balloon burst? 72

Is light made of particles? 76
What colour is light? 78
Can you see rainbows in bubbles?.. 80
When is light invisible? 82
Why is the sky blue? 84
How fast is light? 86
Can you travel at the speed of light? ... 88

In the BEGINNING

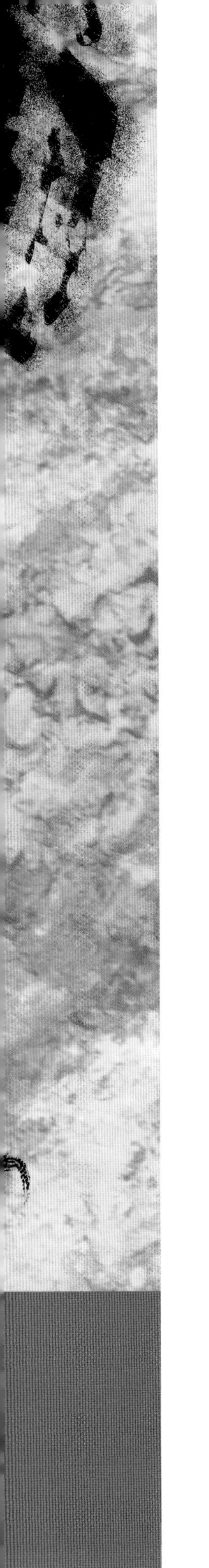

People have been using physics since... well, since people began.

It came in very handy for pushing mammoths off cliffs, lighting fires, throwing spears, and pushing more mammoths off cliffs. So we human beings have always been pretty good at using physics.

But we weren't so good at ***understanding*** *why things worked the way they did.*

Why did a spear fly in a curve when we threw it? How did fire burn our hands and cook our food? And why did the mammoth fall off the cliff? We only really began to find answers when we started doing experiments and measuring things. *And to understand how that got going, we have to travel back in time about 3000 years...*

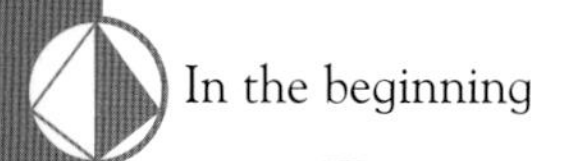

Greece is the *word*

Since the dawn of time, people relied on superstitions and *myths* to explain how the world works. But about 3000 years ago, in Greece, things changed. People stopped trusting their ancient beliefs and decided to think everything through from scratch. It was almost the beginning of science (but not quite).

Amber

The word "electricity" comes from *elektron,* the Greek for amber.

Amazing amber

The Greeks were philosophers (thinkers) rather than true scientists – they had great ideas, but they rarely carried out experiments to check them. Even so, they did make scientific discoveries. They'd discovered **static electricity** by 600 BCE. They knew that rubbing amber (a kind of stone) with wool made it pull feathers as if by magic.

600 BCE

400 BCE

Magnetic soles

According to legend, a Greek shepherd called Magnes discovered the force of **magnetism** when his feet stuck to a mountain, the iron nails in his sandals pulled by a magnetic rock called lodestone. The Greeks thought lodestone contained a "soul" that pulled iron towards it.

What's the matter?

The Greeks came up with the theory that everything is made of **atoms** – particles so tiny that nothing could be smaller. They had no evidence for this bold theory, it was really just a lucky guess. They thought the shapes of atoms might explain their properties, so they gave fire sharp, spiky atoms and made water atoms more rounded.

Fire

Water

Air

Earth

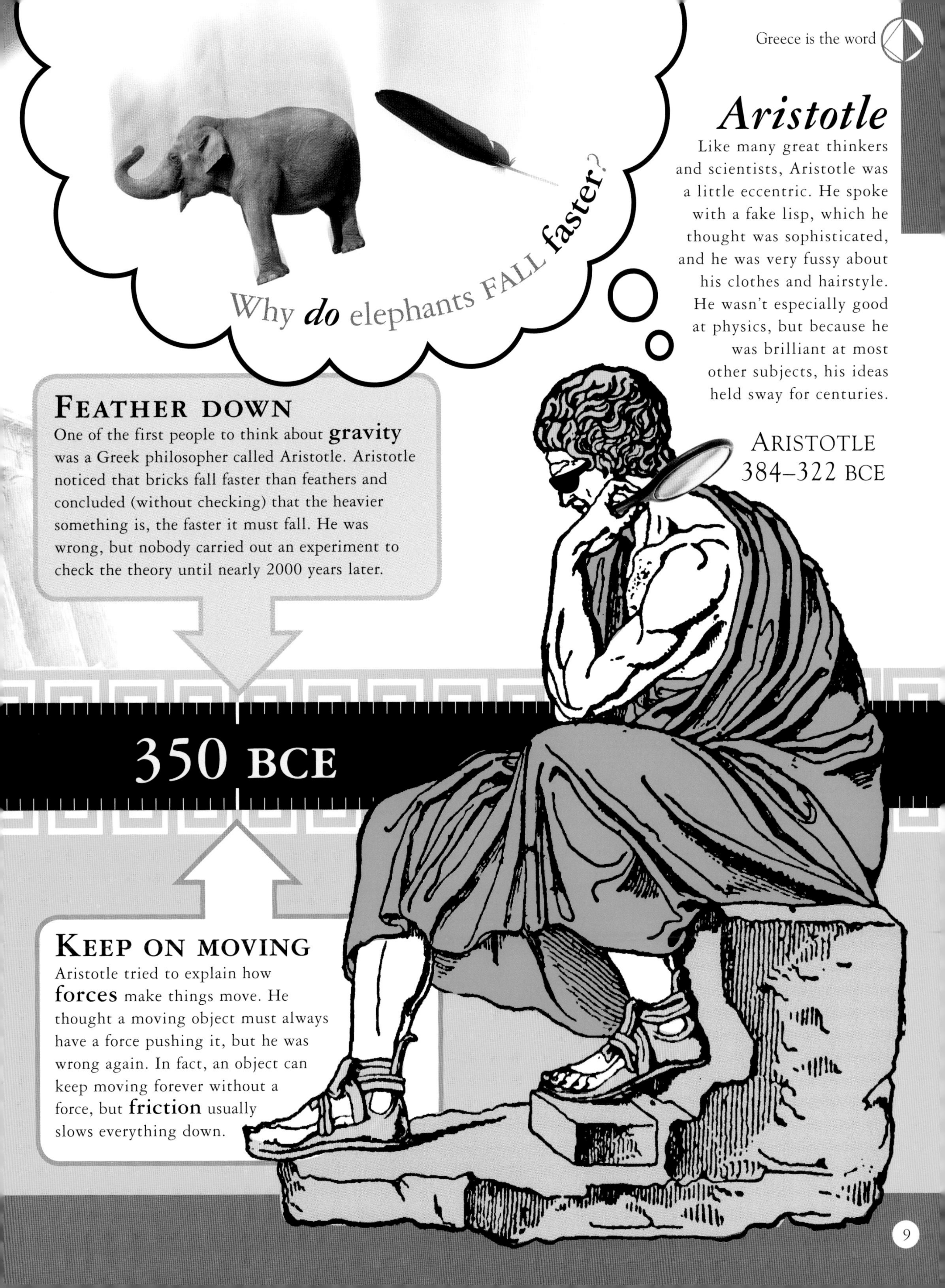

Aristotle

Like many great thinkers and scientists, Aristotle was a little eccentric. He spoke with a fake lisp, which he thought was sophisticated, and he was very fussy about his clothes and hairstyle. He wasn't especially good at physics, but because he was brilliant at most other subjects, his ideas held sway for centuries.

ARISTOTLE
384–322 BCE

FEATHER DOWN

One of the first people to think about **gravity** was a Greek philosopher called Aristotle. Aristotle noticed that bricks fall faster than feathers and concluded (without checking) that the heavier something is, the faster it must fall. He was wrong, but nobody carried out an experiment to check the theory until nearly 2000 years later.

350 BCE

KEEP ON MOVING

Aristotle tried to explain how **forces** make things move. He thought a moving object must always have a force pushing it, but he was wrong again. In fact, an object can keep moving forever without a force, but **friction** usually slows everything down.

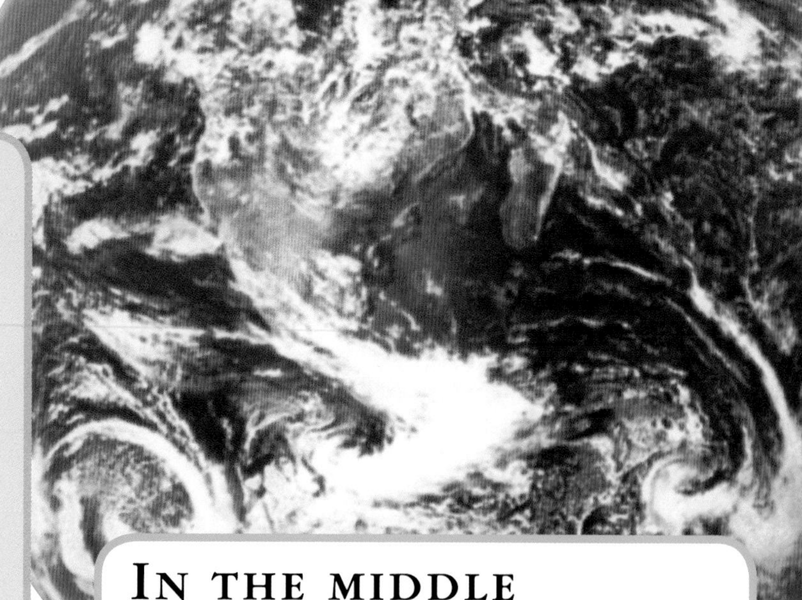

EUREKA!

The cleverest of the Greeks was a man called Archimedes, who once leapt out of his bath and ran **naked** down the street shrieking "Eureka!" after solving a puzzle. The king had asked him to examine a new crown and find out, without cutting it, if it was made of solid gold. The solution came to Archimedes in the bath. He could measure the crown's **volume** by placing it in water and seeing how much the level rose. If an equal weight of gold had the same volume, the crown must be real. As it turned out, the crown was fake and the goldsmith was executed.

IN THE MIDDLE

Most people used to think Earth is flat, but the clever Greeks not only realized Earth is round but figured out its size by measuring shadows in different places. But they didn't know Earth is *turning*, so they naturally thought the Sun and stars cross the sky because they're moving around us. Because the Greeks got this wrong, they thought Earth was the **middle of the Universe**, and this idea stuck for many centuries.

250 BCE

240 BCE

WEAPONS OF WAR

Archimedes was a brilliant inventor. He figured out how levers magnify forces and then used the principle of the lever to build war machines for fighting the Romans. One was an enormous wooden crane with a hook dangling from a rope. It could pick up ships just offshore and roll them over or smash them on the rocks, killing everyone on board.

With a long enough lever I can lift anything!

ARCHIMEDES
287–212 BCE

Ancient Greeks thought planets and stars

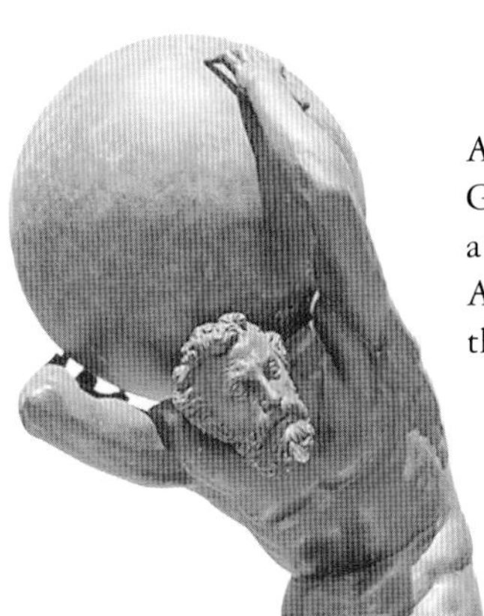

According to Greek legends, a god called Atlas held up the Universe.

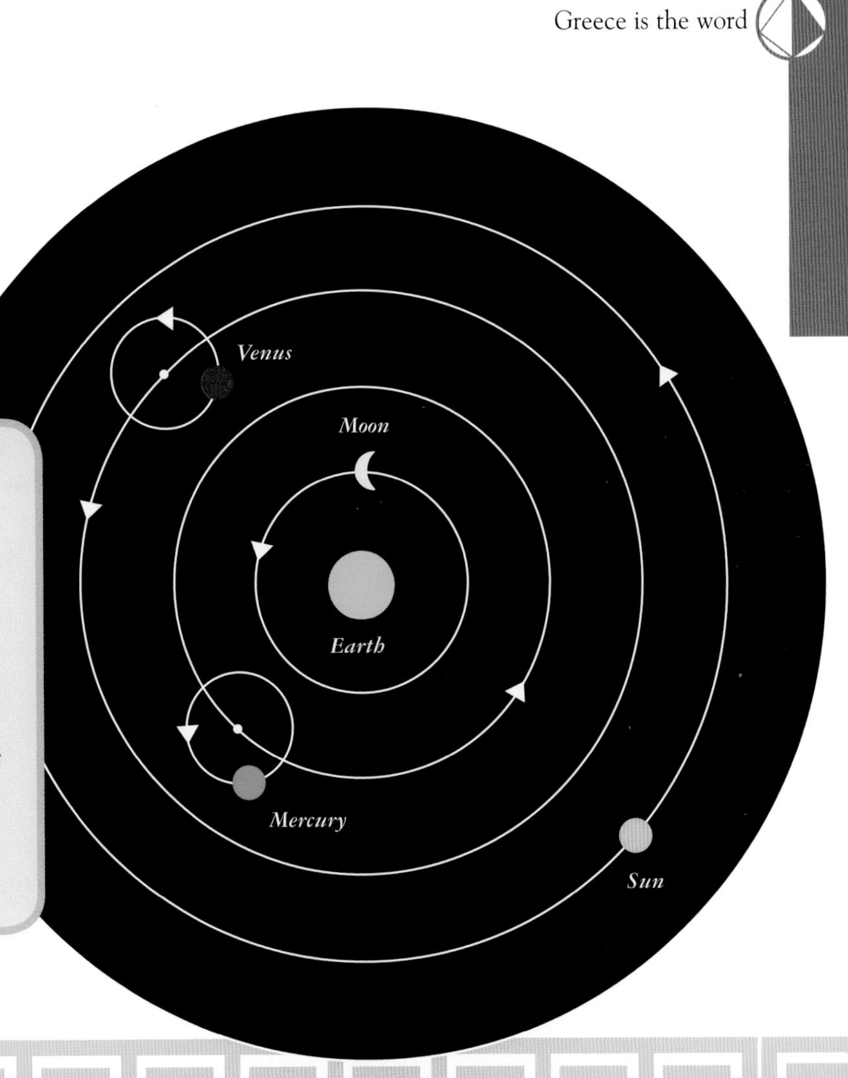

Heavenly spheres

Since the Sun, stars, and planets seemed to go around Earth, the Greeks thought the Universe was based on circles. They thought each planet was attached to a gigantic glass sphere that revolved around Earth, with the stars all on one whopping great sphere outside the planets. The Greeks found they could use this theory to predict where planets would rise and fall, though they had to add fiddly little spheres to the large spheres to make it work. In fact, it worked so well that people carried on believing in it for nearly 2000 years.

150 BCE **50 CE**

Greek geek

One of the last of the Greek philosophers was an inventor called Hero. He built all manner of strange contraptions, including mechanical birds that sang, a mileometer for chariots, a machine gun, and the world's first coin-operated vending machine. Hero realized that air is a substance and discovered he could compress it. This led him to believe it must be made of atoms.

were all stuck to colossal glass balls

The DARK AGES

After the Greek era ended, people turned back to myths, magic, and religion. For more than a thousand years, superstition ruled the world, apart from a few glimmerings of scientific progress...

500 CE

700 CE

It's a kind of magic

The apparently magical abilities of **magnets** led to all sorts of myths in the Dark Ages. People thought a magnet held to the head would cure a person of sadness. They thought garlic and diamonds destroyed the power of magnets and that dipping a magnet in fresh goat's blood restored it to full strength.

Pointing the way

While Europeans were busy rubbing their magnets with garlic and blood, the Chinese were rubbing theirs with iron needles, which became magnetic too. When one of these needles was hung on a thread, it pointed north. The Chinese had invented the pocket **compass**.

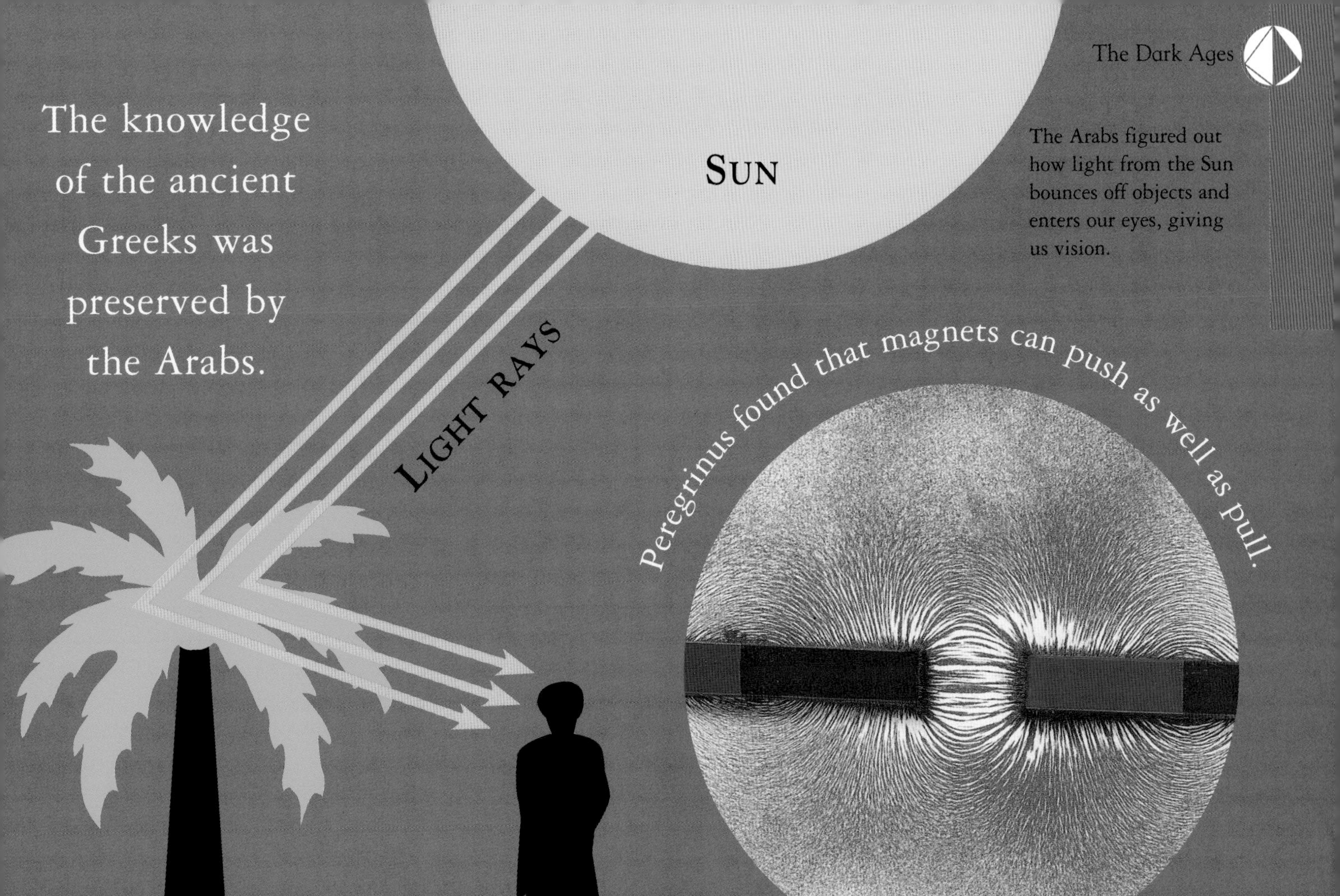

1000 CE

1300 CE

Seeing the light

The teachings of ancient Greece were lost from Europe but preserved in the Arab world, where the spirit of science survived. The Greeks had thought we see because our eyes emit invisible rays that bounce back. In 1000 CE, an Egyptian called Alhazen realized the truth: **light** from the Sun or flames bounces off objects and only then enters our eyes.

Late Chinese compass

Poles apart

Back in Europe, a French man, Peregrinus, tried to separate the opposite poles of magnets by snapping them in half. He found, to his amazement, that each half always became a complete magnet with its own two poles, no matter how many times he divided it.

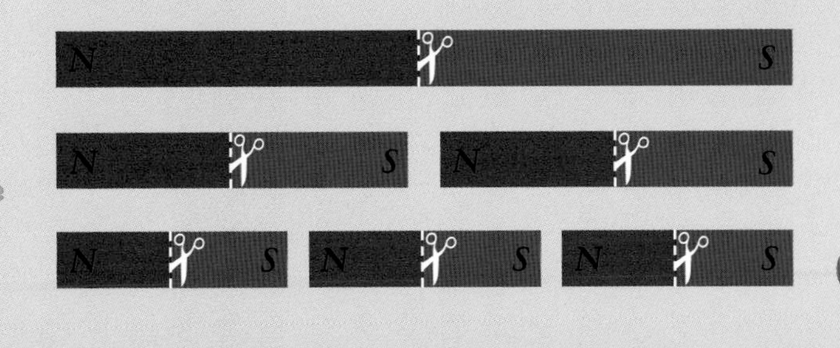

Into *the light*

About 500 years ago

in Europe, there was an amazing change in the way people thought. As in ancient Greece, people began to question their old religious and superstitious ideas about the world, but this time they did something else too. They did **experiments** to test which ideas were right. It was the beginning of science, and it changed the world forever.

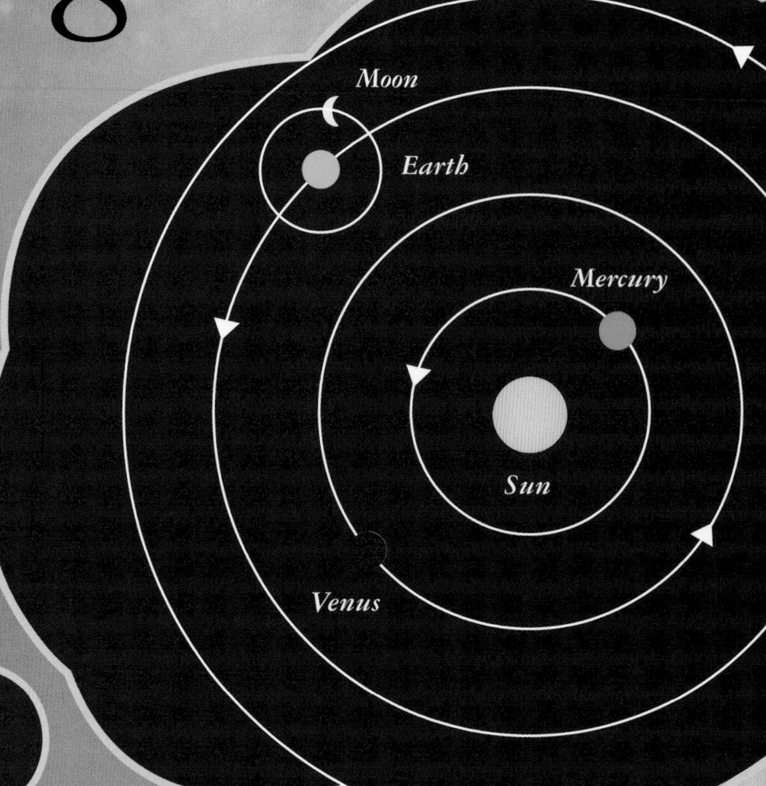

Nicolas Copernicus
(1473–1543)

1507

A brilliant idea

Since Greek times, people had thought Earth sits motionless in the middle of the Universe, with the Sun and planets all whirling around us. In 1507, a Polish astronomer called Nicolas Copernicus noticed it was much easier to predict the motions of planets if you supposed the Sun was in the middle with Earth travelling around it, rather than vice versa. When other astronomers put Copernicus's theory to the test, it worked well. But there was a big problem...

Could the ***whole world*** really be spinning?

Copernicus put the Sun in the middle of his model of the Universe, even though this meant Earth must be flying through space.

The way science works

The difference between science and other ways of thinking is that scientists carry out experiments to check if a theory (an idea) is right. William Gilbert of England was the first person to study magnets scientifically. Instead of simply accepting the old story about garlic ruining magnets, he rubbed garlic on magnets and tested them – it made no difference. Gilbert also noticed that magnetic needles point slightly to the ground and formed a theory that Earth is a magnet. He was right.

1543

1580

A dangerous idea

Religious people hated Copernicus's idea because they believed God created the Universe with Earth in the middle. Copernicus put the theory in a book but, afraid of offending the Church, waited until he was on his deathbed before publishing it and added a flattering dedication to the Pope.

A mad idea?

There was another reason Copernicus's idea was hard to believe. If it were true, it meant the Sun's motion across the sky was just an illusion caused by Earth spinning round at ferocious speed. People thought that was impossible. If Earth were spinning, they thought, birds and clouds would be left behind and buildings would fall over. This puzzle was solved by the next person in our story...

Copernicus's book

Galileo's WORLD

The world's ***first proper scientist*** was an Italian called Galileo. His ingenious experiments showed that the Greeks hadn't just been wrong about space – they were wrong about gravity and motion too. Galileo's discoveries were the start of physics, but they got him into *deep trouble*.

Leaning Tower of Pisa

TOGETHER WE FALL

Legend has it that Galileo dropped balls of different weights off the Leaning Tower of Pisa to show they hit the ground together. In fact, he probably just rolled balls down ramps, but he did disprove the old Greek idea that heavy objects fall proportionately faster than light ones. As Galileo found, all objects are pulled down by gravity at the same rate, unless they are so light or fluffy that the air can slow them down.

This *is where physics* **REALLY** *starts!*

1590

Galileo figured out the precise curve of a ball in flight by separating it into two parts: a constant horizontal speed and a changing vertical speed. This solved an age-old problem: how to work out the path of a cannonball.

ON A ROLL

As well as rolling balls down ramps, Galileo rolled them over level surfaces and flung them around his room. He measured their tracks with enormous care and timed their every movement, using his pulse as a stopwatch (since clocks hadn't yet been invented). And he made an amazing discovery. The falling balls went faster and faster, with the pull of gravity speeding them up continually, but a ball rolling along a flat surface kept a constant speed ***without any force***. Galileo had discovered **inertia**: the tendency of an object to either keep going or to stay still unless a force is acting on it.

Galileo the astronomer

Galileo was brilliant at everything he did. In 1609 he heard about the invention of the telescope and set about making his own. It was the best in the world. Using it, he discovered mountains on the Moon, moons around Jupiter, and fantastic numbers of tiny stars, which hinted at the vast size of the Universe. What he saw convinced him Copernicus was right about Earth going round the Sun, and he said so in a book in 1632. But religious leaders weren't impressed. They banned the book and had Galileo imprisoned.

Galileo's fish

Galileo's discovery of inertia led him to realize that there's actually no difference between moving at a steady speed and being still. He put it like this in a book: imagine a fish swimming in a tank on a ship that's sailing on the sea. As far as the fish is concerned, the ship might as well be still. The fish doesn't get pushed to the back of the tank by the ship's motion – it swims normally. For just the same reason, clouds and birds don't get left behind and buildings don't fall over as Earth spins round – they all keep moving around with everything else, because they have inertia. Galileo had solved the most puzzling problem with the theory that Earth goes round the Sun.

Newton's *Universe*

Galileo died in 1642, a prisoner in his home. Fortunately, on Christmas Day of the same year, an even **cleverer** scientist was born in England. He was bad tempered, miserable, and peculiar, but he was also a **genius**. His name was **Isaac Newton**.

In 1666 Newton went to stay on his mother's farm to avoid the deadly plague that was sweeping through English towns. One day in the orchard, he saw an apple falling from a tree. He began to wonder if **gravity**, the force that pulled the apple to the ground, might also hold the Moon in its grip.

Hmm... *I can sense* **GRAVITY** ***in action!***

1666 *and beyond...*

Before Newton, nobody understood what kept the Moon flying around Earth, or the planets flying around the Sun. People had always thought they were moved by the gods or pushed along by some invisible force, but Newton **figured out the true answer**.

Isaac Newton (1642–1727)

Newton realized the Universe is held together

Galileo had already worked out exactly why cannonballs fly along curved paths. Newton realized that the **Moon is like a gigantic cannonball** that is flying too fast to fall back to the ground. To explain this theory, he drew a diagram showing what would happen if you fired a cannonball faster and faster into space from the top of a mountain. At first it would fall back quickly. As it got faster, its curve would get gentler. Eventually it would fly so fast that its path would be less curved than Earth itself. It would carry on flying, forever falling and never landing. In other words, it would be trapped **in orbit**.

Newton realized there was no need for a force to push the Moon along. As Galileo had already discovered, a moving object keeps going because it has **inertia**. Newton saw that inertia keeps objects moving in a straight line unless a force pushes them away. The Moon "wants" to fly in a straight line, but the force of Earth's gravity keeps tugging it back, making it fly in a curve.

Newton's drawing

Newton also realized that planets are "falling" around the Sun in just the same way, trapped in orbit by the Sun's immense gravity. He worked out the precise shapes and speeds of their orbits with maths, a fantastically complicated task that took years to complete, and forced him to invent a whole new branch of maths called calculus. But it was worth it. He'd succeeded in figuring out both gravity and three "laws of motion" that govern how *everything in the Universe* moves, from atoms to galaxies. It was the **greatest scientific discovery ever.**

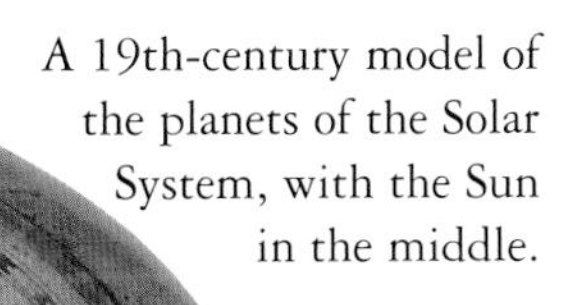

A 19th-century model of the planets of the Solar System, with the Sun in the middle.

Nasty Newton

Isaac Newton hated people and spent most of his life alone. At school he was a strange boy with few friends and spent his free time making strange contraptions, such as a kind of windmill powered by mice. As an adult he was always quarrelling. He accused other scientists of stealing his ideas, and he once told his mother and stepfather that he was going to burn down their house and kill them. Even his scientific work was sometimes odd, at least by today's standards. He wasted years trying to find a recipe for gold (which is impossible), and he used the Bible to calculate the year of creation as 3500 BCE – out by 4.5 billion years.

Can you *feel* the FORCE?

Physics started with the idea of forces. So what exactly is a force and how does it work? Is it something you can capture in a bottle or examine under a microscope? Does it glow in the dark or fizz if you drop it in water? Well... no.

A force isn't really a physical thing – it's more of an idea.

And a pretty simple idea at that, because a force is just a ***push or a pull***. Understanding forces can answer all sorts of questions, like why rollercoasters push your stomach up into your chest, why cats can jump off skyscrapers and live to tell the tale, and why a bicycle can accelerate faster than a sports car.

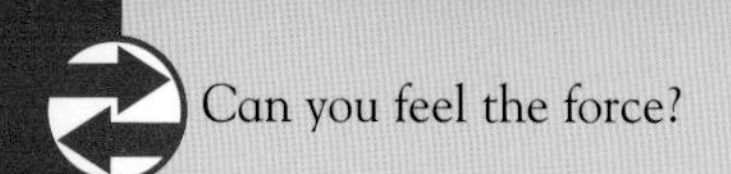

What is a FORCE?

Forget *Star Wars* – a force is not some mysterious, invisible energy field that pervades the whole Universe. (Though mysterious, invisible energy fields that pervade the whole Universe do exist.) It's actually something very simple: just a push or a pull.

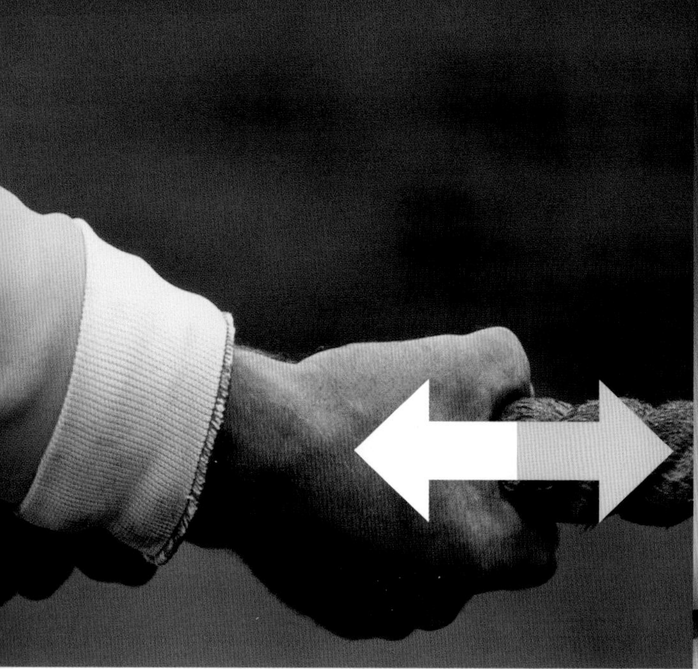

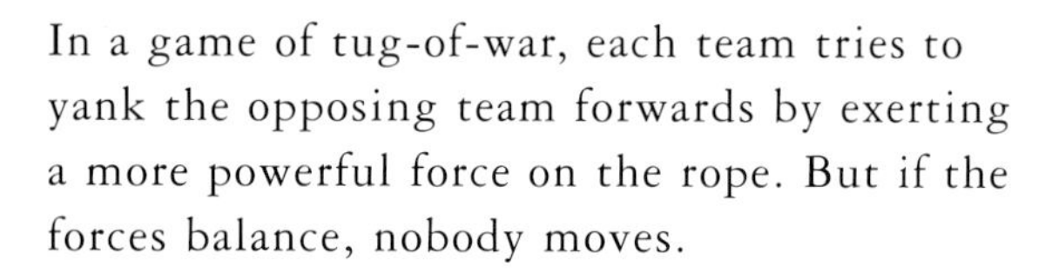

In a game of tug-of-war, each team tries to yank the opposing team forwards by exerting a more powerful force on the rope. But if the forces balance, nobody moves.

These sumo wrestlers are both leaning forwards as they try to push their opponent away with their body weight. Since they weigh the same amount, the forces balance and neither wrestler moves.

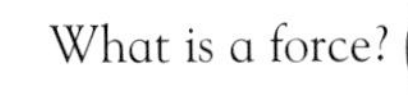

Some forces can *push and pull* objects without even touching them.

There are forces acting on you *all the time*, whether you know it or not. While you read this, gravity is pulling you down, the ground is pushing you up, air is squeezing you from every direction, and the matter in your body is pushing back at the air. When forces balance, they cancel out and you don't notice them. But when forces *don't* balance, things happen...

Gravity is trying to pull this house back to the ground, but its enormous weight is balanced by an upward supporting force from the ground, transmitted through the truck.

As the boxer's fist delivers a powerful force to the punch bag, it tries to swing back under the force of its own weight. But the forces don't balance, so the punch bag swings to the right.

It's *the* LAW!

While he was figuring out how gravity governs the movements of the planets, **Isaac Newton** discovered three simple laws that describe how forces make things move. These "laws of motion" laid the foundation for the whole science of physics, and they work for just about everything, from fleas to footballs and from atoms to planets.

1 An object that isn't being pushed or pulled by a force either *stays still* or *keeps moving* in a straight line at a constant speed.

2 *Forces* make things accelerate. The bigger the force, and the lighter the object, the *greater the acceleration*.

3 Every action has an *equal* and *opposite* reaction.

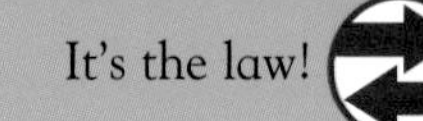

... in other words

if you let go of a moving shopping trolley, it keeps going in a straight line until it crashes into something.

This law is all about ***inertia***. It's common sense that an object stays still until something pushes it, but what about the second half of the law? In our everyday experience, moving objects don't keep on going forever at a constant speed – they grind to a halt. That's because ***friction*** and other forces are slowing them down. But if you take away friction – by wearing ice skates, sitting in a shopping trolley, or going into space – Newton's first law works much better.

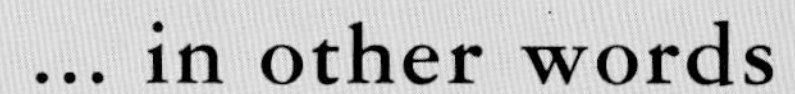

... in other words

it's much easier to make a lightweight racing bike speed up than a ten-ton lorry.

This law tells you exactly what happens when you give something a push. In everyday language, "acceleration" means speeding up, and that's what happens if you push something forwards. The harder you push the pedals on a bike, the faster you go. And the less you and your bike weigh, the easier it is to speed up. In physics, acceleration doesn't just mean speeding up – it can mean ***any change*** from either being still or travelling in a straight line at a steady speed. So when you pull on your brakes to slow down, the force of friction gives you ***negative acceleration*** (you decelerate).

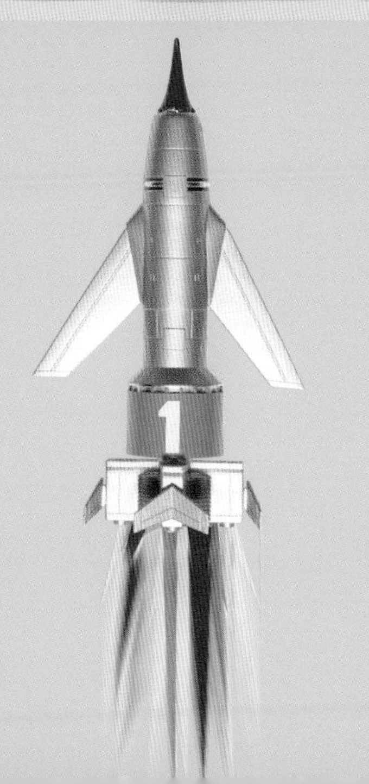

... in other words

when a rocket pushes burning gas out of its engine, the gas pushes back on the rocket and lifts it into space.

Newton realized that forces (which he called "actions") ***always happen in pairs***. If one object pushes another, the second object pushes back with the same force. The forces are equal, but their effects may not be. If you throw a ball, the ball pushes back on your hand, but only the ball flies away. If you push the ground quickly with your feet, the ground pushes back and flings you into the air – you jump. The force from your feet also moves the whole of planet Earth down, but only by a tiny amount!

Can you *do* PHYSICS

To see how Newton's laws work, all you have to do is get on a bike and go for a ride...

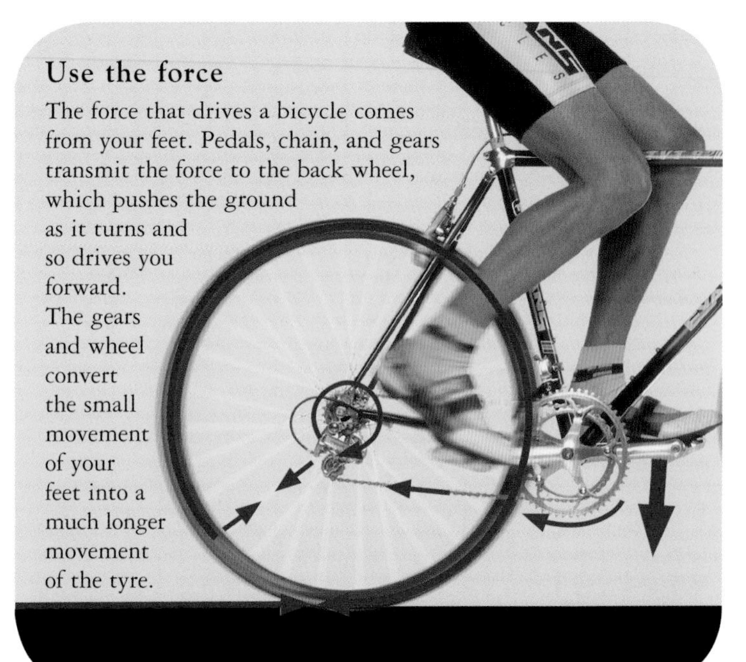

Use the force

The force that drives a bicycle comes from your feet. Pedals, chain, and gears transmit the force to the back wheel, which pushes the ground as it turns and so drives you forward. The gears and wheel convert the small movement of your feet into a much longer movement of the tyre.

1

Setting off

Before you set off, your bike is standing still because there's no force acting on it. That's Newton's first law at work. When you get on and start pedalling, you apply a force and the bike accelerates. That's Newton's second law. If the bike is very light, you'll accelerate quickly. That's also Newton's second law.

2

Downhill

You start going down a hill. Now another force is making you accelerate: gravity. You go faster and faster without having to pedal. In fact, you're going a bit too fast so you apply the brakes sharply. They rub the wheels, using the force of friction to slow you down.

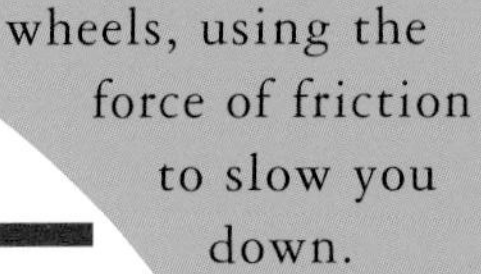

3

Wipeout

Yikes! You must have pulled on the brakes too hard. The bike stopped, but you didn't. Inertia kept your body moving in a straight line at a constant speed and you flew over the handlebars. Blame Newton's first law. You get back on the bike and coast downhill again.

4

Freewheeling

When you reach the bottom of the hill you stop accelerating, but you don't have to pedal just yet – you can freewheel. Your bike keeps moving at a steady speed because of Newton's first law. Now your body's inertia is helping you.

What a drag

The main thing holding you back as you cycle along the flat is "drag" – a complicated kind of friction caused by air. Low posture, smooth clothes, and a streamlined helmet can all help to reduce it.

... on a *bike?*

On the flat

As you freewheel along the flat, you lose speed. But Newton's first law says moving objects keep going at a steady speed, unless there's a force. So there must be a force. It's friction again, but this time it's caused mostly by the air pushing against you. To maintain your speed, you have to pedal just enough to balance the force of friction.

5

Uphill

Now gravity is pulling the bike again, but this time it's working against you. Thanks to Newton's second law, gravity makes you decelerate. Gravity is more powerful than friction with the air, so you have to pedal much harder to overcome it. Cycling uphill is hard!

6

Turning

The road bends, so you steer the handlebars to make a turn. Newton's first law says you should keep going straight unless there's a force, so what's the force? It's actually friction again. When you turn, friction between your wheels and the road becomes one-sided, pushing you away from a straight line.

7

8

Get a grip

So where does Newton's *third law* come in? The third law explains why a bike works at all. As you force the back wheel round, its tyre grips the ground and pushes it. The ground pushes back equally, driving you forward.

What *causes* FRICTION?

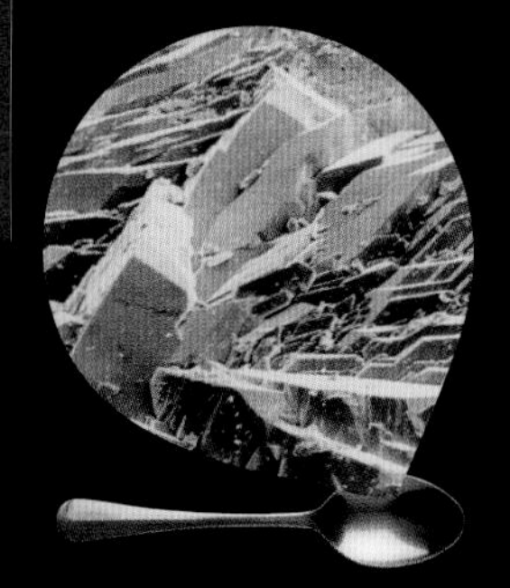

No matter how smooth an object might look, in reality it's covered with millions of tiny jagged bumps and dents. When two objects rub together, the tiny bumps snag each other and slow the objects down. This slowing force is ***friction***, and it's the enemy of motion. But sometimes it's a good thing, because friction also gives you **grip**.

How to beat friction

One way to beat friction is to cover moving objects with a slippery liquid – a lubricant. If all the moving parts on a bicycle are well lubricated with oil, the bicycle will be more efficient because less energy is wasted in friction. Even solids can act as lubricants. Snow and sand are both slippery enough to ski down because they are made of small grains that slide past each other.

GRIP

Traction control

Walking would be impossible without friction. If there were no friction between your feet and the ground, your shoes wouldn't grip and you'd slip with every step. Shoes and tyres with a deeply patterned tread give maximum grip, or "traction", ideal for slippery surfaces.

Fact or friction

How do flies walk up walls?

Flies can run up walls and scamper over ceilings, so why don't they fall off? Friction is part of the answer, and their feet also have tiny hooks for hanging onto crevices. But the real secret of their sticky feet is soft pads covered with hundreds of tiny, damp hairs. The hairs mould to the shape of any surface and form a sticky seal.

NOW TRY THIS!

To prove how powerful friction can be, interleave the pages of two paperback books. Challenge a friend to pull them apart with strength alone. They won't be able to!

Two types of friction

There are two types of friction: static friction and sliding friction. Static friction is much more powerful and makes it hard to budge an object that isn't moving, like a heavy box on the floor. Once the box *is* moving, you can push it fairly easily, because only the weaker force of sliding friction is trying to slow it down.

It won't budge!

Easy!

STATIC FRICTION

SLIDING FRICTION

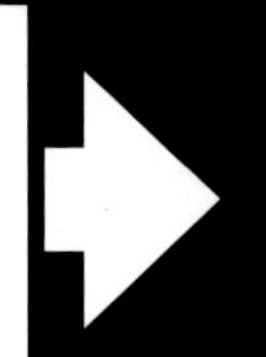

What happens when you skid?

Cars and bikes normally grip the road with **static friction**. When a wheel goes into a skid, however, the point of contact slides over the ground, so the only force gripping the surface is **sliding friction**. Because sliding friction is weaker than static friction, cars and bikes have much less grip when they skid and are harder to control.

Balls of steel

Bearings do the same job as lubricants, reducing friction between moving parts. Bicycles have bearings in the wheels, pedals, and the tube under the handlebars.

SLIP

Get a grip

Ever wondered what your fingerprints are actually *for*? The answer is friction. They work the same way as the soles of shoes or the tread of a tyre. The ridges of skin also have tiny pores that ooze water and oil to make them stickier.

Fact or friction

Antilock braking

Used properly, bike brakes should always create sliding friction. If you brake too hard, static friction takes over and the wheel locks, causing a skid. On icy roads the tyres have much less grip and can skid very easily. If you ride on ice, brake little and often and use the back brake only.

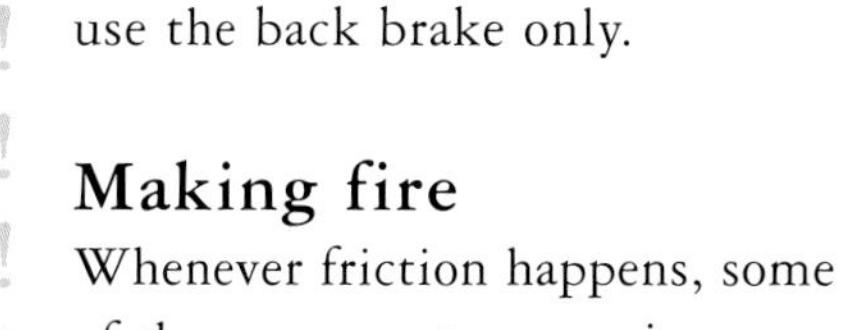

Making fire

Whenever friction happens, some of the movement energy in an object turns into heat energy. (Rub your hands together as hard as you can for proof.) If there's enough friction, the heat can trigger a fire. People first realized this about half a million years ago when someone had the brilliant idea of rubbing two sticks together.

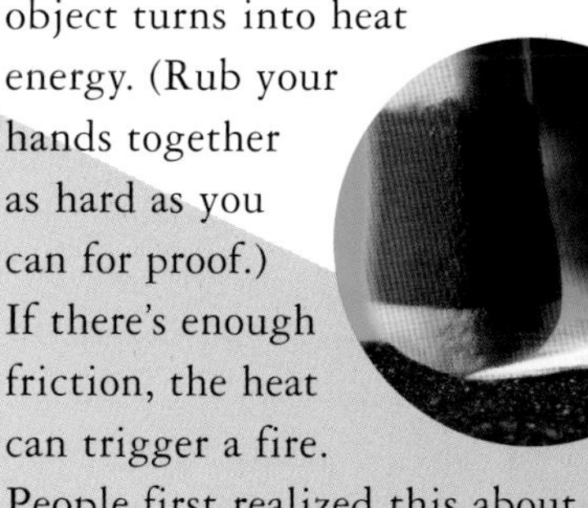

velocity = distance ÷ time

To work out your speed, divide how far you travel by how long it takes. So if you run 24 km (or miles) in two hours, your speed is 12 kph (or mph). Physicists usually talk about ***velocity*** rather than speed. Velocity is your speed in a particular direction. If you run north and turn east without slowing, your velocity north drops to zero but your speed doesn't change.

How fast can you go on Earth?

How fast you can go depends on where you are. Travelling in water is slow because water creates lots of friction. In fact, the world's fastest submarine is slower than a bus. On land, friction with the ground can hold you back. The first land vehicle to exceed the 1224 kph (760 mph) speed of sound is a British car named *Thrust SSC*. Travelling faster than sound is easier in air, where jets routinely break the sound barrier, causing weird cloud formations and sonic booms.

How fast can you go in space?

In space, where there's no air to cause friction, spacecraft can reach thousands of kilometres per hour. But it feels just the same as being still. You only feel the effect of motion in space if you *accelerate*. And no matter how powerful your rocket, you can't exceed 1 billion kph (670 million mph) – the speed of light. It's the ultimate speed limit.

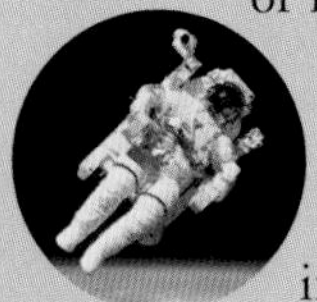

How *FAST*

On Earth, moving objects don't usually keep going in a straight line at a constant speed – they're always speeding up,

How fast is the fastest... space probe?

... manned spacecraft?

... car?

... production car?

... cyclist?

... caravan tow? 228 kph (142 mph)

... land animal? 100 kph (62 mph) cheetah

... tank? 82 kph (50 mph) *Scorpion Peacekeeper*

... submarine? 74 kph (46 mph) *Russian Alpha*

... insect? 56 kph (35 mph) dragonfly

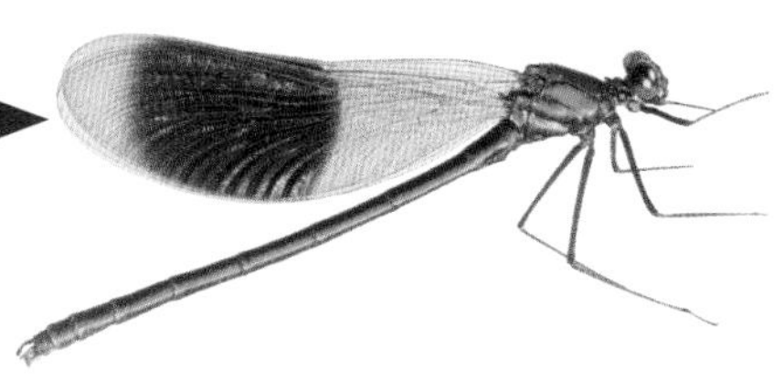

... person? 43 kph (27 mph)

can *you* GO?

slowing down, and changing direction. To understand what they're doing, you need to know their *velocity* and *acceleration*.

252,800 kph (157,083 mph)
Helios spacecraft

40,000 kph (24,854 mph)
Apollo spacecraft

1228 kph (763 mph)
Thrust SSC

431 kph (268 mph)
Bugatti Veyron Super Sport

269 kph
(167 mph)*

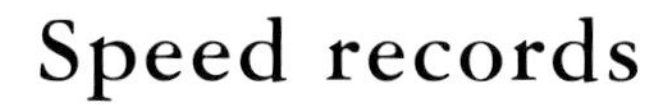

Speed records

An object's top speed depends on three things: its power, its weight, and the amount of friction it has to overcome. With the right power-to-weight ratio, even a humble dragonfly can move faster than the world's fastest Olympic sprinter.

* This unbelievable cycling record was achieved by riding in the slipstream of a dragster.

How fast are you *really* going?

Speed is relative. You may feel as if you're sitting quite still as you read this, but think for a moment. Planet Earth is spinning round, which means you're actually travelling east at up to 1600 kph (1000 mph). And Earth isn't just spinning – it's flying around the Sun at more than 110,000 kph (70,000 mph), which means you are too. Then again, the Sun and the solar system are hurtling through space at 2 million kph (1.3 million mph). So how fast are you *really* going? There's no right answer – it just depends on your point of view.

acceleration = change in velocity ÷ time

Acceleration is how quickly your velocity is changing. It's a tricky concept to understand. It doesn't just mean speeding up – it means ***any change in velocity***. Speeding up and slowing down both count as acceleration, and so does changing direction, because velocity has a direction.

What's the force you feel?

Unlike speed, acceleration is something you really ***feel***. When a powerful car accelerates, you feel pushed into the back of the seat by invisible hands. Likewise, you feel pushed forwards when you decelerate, and you feel pushed sideways when you turn. This pushing feeling isn't a real force – it's just your inertia trying to keep you moving in a straight line at a steady speed. But it feels just like the force of gravity, so we call it ***g* force**. When you ride in a lift, *g* force acts in the same direction as real gravity, making you feel heavier or lighter than you really are.

GOING UP

g force makes you feel heavier

GOING DOWN

g force makes you feel lighter

G FORCE

The ***exhilarating feelings*** you get on a rollercoaster are caused by *g* force. With every twist, turn, rise, and fall, the train is changing speed. And every time it changes speed or direction, *g* force pushes your body.

G force acts on every part of your body, including your internal organs, which are only loosely connected together. As you plunge downhill, your stomach and intestines lurch upwards and press against the base of your lungs. And as the train turns up at the bottom of a valley, all your internal organs are squished downwards.

At the bottom of a hill on a big rollercoaster, *g* force ***triples*** your body weight

At the crest of a rollercoaster hill you experience ***negative g force***. It acts in the opposite direction to gravity, cancelling your weight and making you float upwards. The front and back seats get the most powerful negative *g*'s because they ride over the crest quickest.

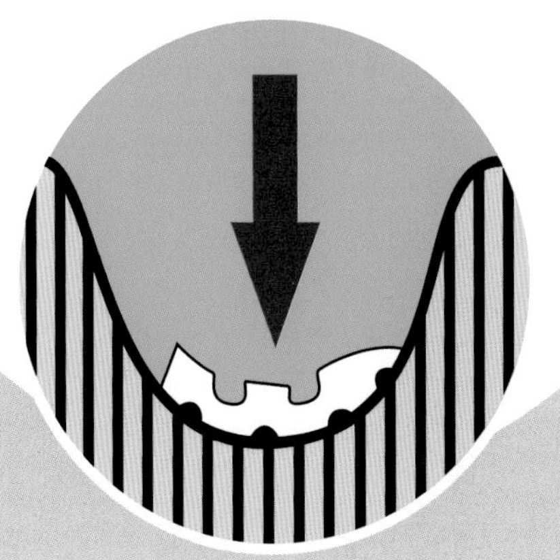

At the bottom of a valley you feel ***positive g force***, which multiplies the force of gravity and presses you into your chair, making you up to three times heavier. The middle seat gets the highest positive *g* force because it passes through the bottom of the valley quickest.

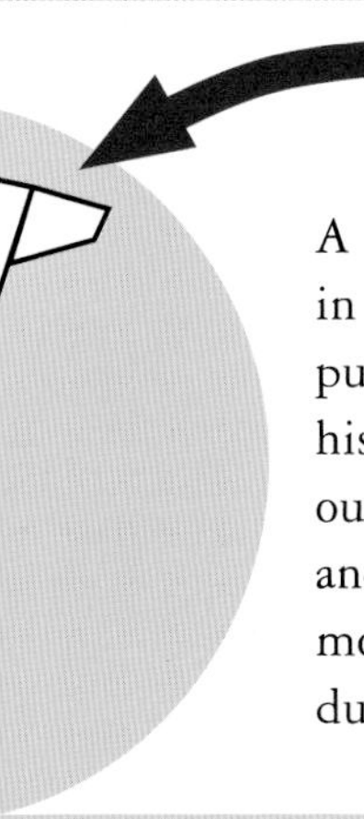

A fighter pilot endures up to 9 *g* in sharp turns. This tremendous force pushes blood out of his head and into his legs, which can make him pass out unless he wears a pressure suit and performs forceful clenching movements with lower body muscles during the turn.

In 1954, US scientist John Paul Stapp subjected himself to 46.2 *g* in the name of science. Strapped into a rocket-powered sled on train tracks, he decelerated from 1017 kph (630 mph) to zero in 1.25 seconds – equivalent to hitting a brick wall at 190 kph (120 mph). He survived (there was no brick wall), but his eyes filled with blood and he was temporarily blinded – a "red out".

Can you take the *g*'s?

The human body can barely withstand negative *g* force as it pushes blood up into your head and can burst blood vessels in your brain. However, people have survived brief exposure to amazing high positive *g*'s.

−3 *g*	The maximum negative *g* force a person can safely withstand.
0 *g*	Weightlessness in space.
1 *g*	Normal gravity.
3 *g*	The maximum *g* force experienced on a big rollercoaster.
4.5 *g*	The maximum *g* force civilian aircraft are designed to take.
5 *g*	Most people black out if subjected to more than 5 *g* for a sustained period.
5.1 *g*	Top dragsters can do 0 to 100 kph (60 mph) in half a second, producing 5.1 *g*.
9 *g*	Fighter pilots are trained to withstand 9 *g* during aerial manoeuvres.
46.2 *g*	The maximum *g* force deliberately endured by a human being.
100 *g*	Exposure to 100 *g*, however brief, is nearly always fatal.
180 *g*	The maximum *g* force a human being has survived.

TYPES OF ENERGY

Potential energy is the energy you store when you lift up a weight, squeeze a spring, or stretch an elastic band.

Chemical energy is trapped in molecules. Food, petrol, and other types of fuel are rich in chemical energy.

Kinetic energy is the energy a moving object has. The faster something is moving, the more kinetic energy it has.

Light is pure energy travelling at an amazing speed. Nearly all the energy we use comes originally from sunlight.

Heat is the energy of atoms and molecules shaking. The hotter something gets, the faster the atoms shake.

Electrical energy is a form of energy that can travel conveniently through wires with almost no waste.

Dark energy is a mysterious form of energy that is making the Universe expand.

Nuclear energy is released when atoms are torn apart or fused together in the Sun, nuclear bombs, and nuclear power stations.

ENERGY

Forces can't happen without ENERGY. Whenever a force pushes or pulls something, energy is making it happen. Without energy, nothing in the Universe would happen.

CONVERTING ENERGY

One of the laws of physics is that energy can *never* be destroyed. It just gets converted from one form to another whenever it's used. The energy you use to ride a bike comes originally from nuclear explosions in the Sun. To get to your bike it had to go through several different forms first.

Energy leaves the Sun in the form of light and heat.

Plants capture light energy and store it as chemical energy.

Maximize your potential

Stored energy is sometimes called potential energy. As you climb the hill at the start of a rollercoaster, your potential energy builds up. When you ride downhill, the potential energy turns into kinetic energy, making you go faster and faster.

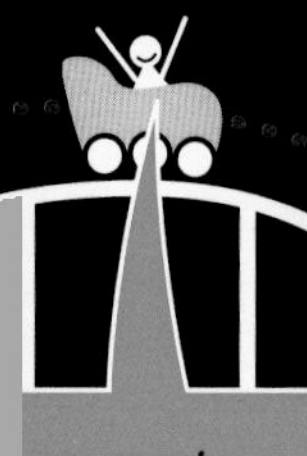

How does energy work?

Energy works a bit like money. You can save it up or you can spend it and make things happen. The energy you save isn't doing anything while it's in storage, but it does have the ***potential*** to make things happen. Using it is like spending money. You get something in return for using it – like light from a torch – but you end up with less in your store.

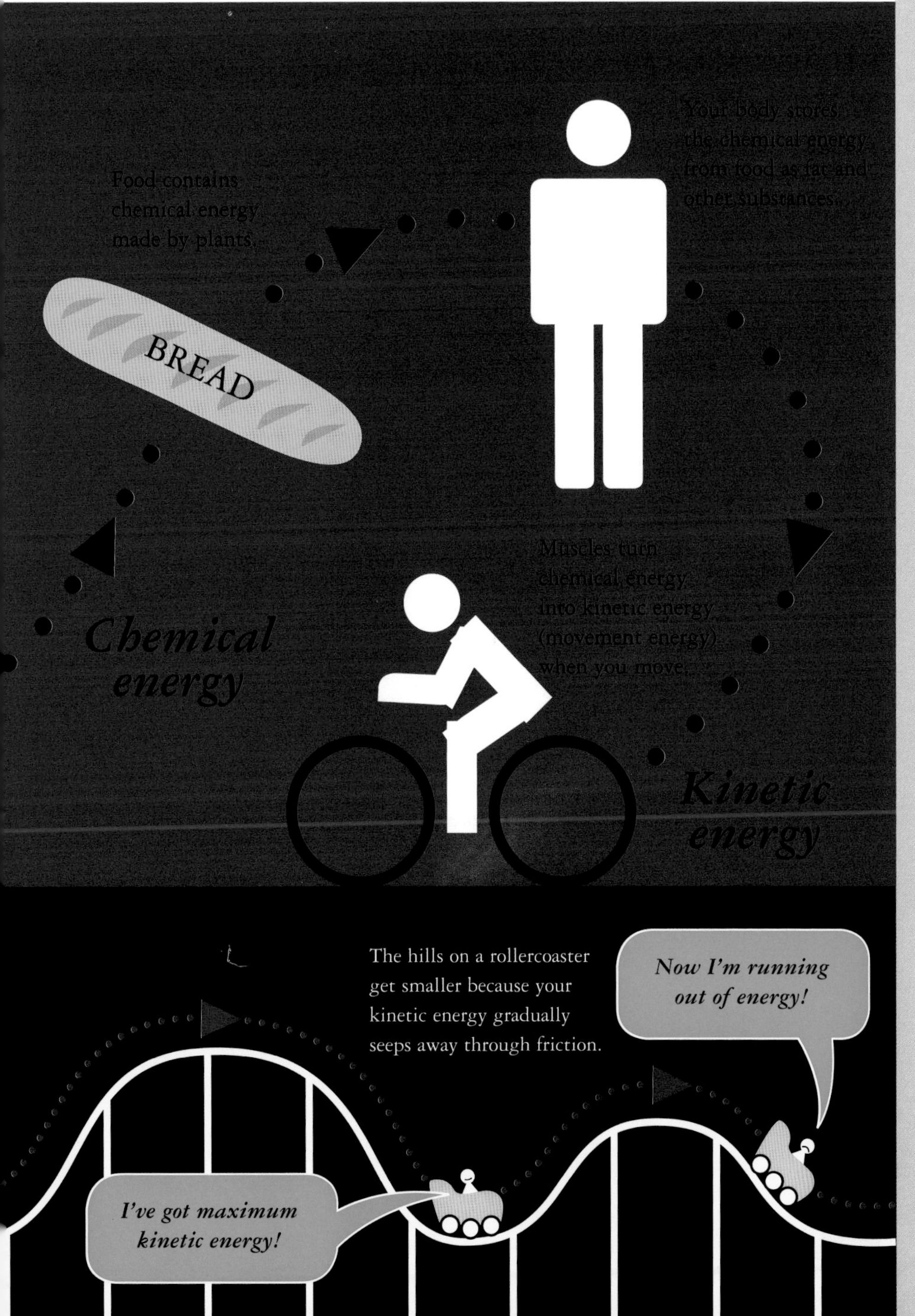

FAQ

How do you measure energy?

Scientists measure energy in joules. One joule is about the amount of energy you need to lift an apple up by 1 m (3 ft). A light bulb uses about 100 joules a second, a person sprinting uses about 1000, and a car uses about 100,000. One slice of cherry pie contains 2,000,000 joules – enough to drive a car for 20 seconds or to lift two million apples by a metre.

Where does energy come from?

Whenever you turn on a light, watch TV, or ride in a car, you're using energy. Most of the energy we use comes from fossil fuels, which are burnt in power stations so the energy can be converted into other forms. Fossil fuels are called nonrenewable energy because they'll run out one day. Other types of energy, such as solar power, are called renewable, because there's an unlimited supply.

$$E = mc^2$$

When an atom bomb explodes, some of the matter turns into pure energy. Albert Einstein's famous equation tells you exactly how much energy (E) you get. To work out the energy in joules, you multiply the amount of matter in kilograms (m) by a ***colossal*** number: the speed of light squared (c^2). That's why atom bombs make such a big bang!

How can you make

Any gadget that magnifies a force is called a MACHINE. Most of the machines you use are so simple you probably don't even realize they're machines. Door handles, hammers, can openers, and wheels all count as machines.

Machines work on a very simple principle: you put in a force at one end and you get a different-size force – usually a bigger one – out the other end. Try pulling a nail out of wood with your fingers. With a claw hammer it's easy, because the hammer ***magnifies the force***.

small force, long distance

The INCLINED PLANE

The simplest kind of machine is the inclined plane, which is just a fancy term for a ramp. It's much easier to carry a heavy load up a long ramp than it is to haul it up vertically. But you have to carry it much farther, and that's the catch.

The LEVER

A lever is a machine with a fixed point – a ***fulcrum*** – that stays still while other parts move. As a result, the lever pivots. Depending on what type of lever it is, the force you put in can be either magnified or shrunk.

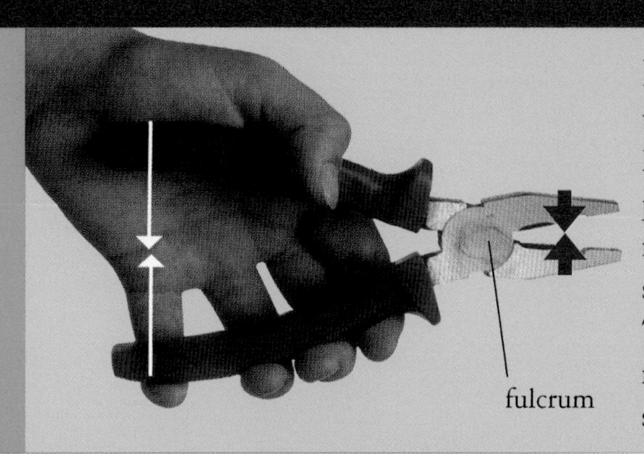

Pliers turn the weak force of your hand into a more powerful gripping force on the other side of the fulcrum. The magnified force moves a shorter distance.

The WHEEL

As well as rolling along the ground to help things move, wheels can magnify or reduce forces in a similar way to levers. The bigger the wheel, the more it changes the size of the forces.

A steering wheel turns the weak force from your hands into a powerful force at the hub. The rear wheel of a bike does the opposite, turning a powerful force from the chain into a weak force at the tyre, but it multiplies the distance.

FORCES bigger?

Machines don't give you something for nothing. There's a catch: the magnified force moves a *shorter distance* than the force you put in. So what you gain in force, you pay in distance.

To put it another way, the energy you put in equals the energy you get out, because you have to obey this law:

energy used = force × distance

large force, short distance

The WEDGE

A wedge works like a ramp, except that it moves. An axe is a wedge. When an axe hits a log, the long downward force is transmitted into much more powerful (but shorter) sideways forces that split the wood.

The SCREW

A screw is simply an inclined plane that's been coiled up. Each turn of a screw pushes it just a tiny bit deeper into the wood, but with much greater force than your hand applies at the other end. A big screwdriver makes the job even easier.

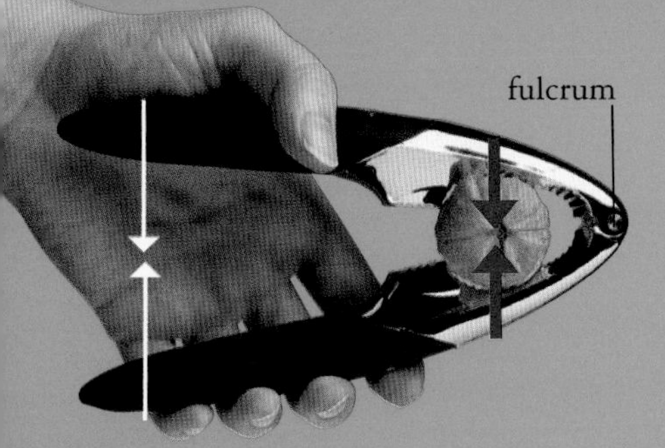

Nutcrackers turn the weak force of your hand into a powerful crunching force on the same side of the fulcrum, but nearer to it.

Chopsticks reduce the force of your fingers but magnify their movements, giving a more delicate grip.

GEARS

A gear is a wheel with teeth that turns a second, similar wheel. In a car engine the gears make direct contact, but in a bicycle they are linked by a chain. If the first wheel has more teeth than the second, the second one turns faster but with less force. If the second wheel has more teeth than the first wheel, it turns slower but with more force.

How do **bicycles** work?

The bicycle is the ***most efficient transport machine*** ever invented, turning the energy you supply into pure forward motion. So how does it work?

To stay balanced, a cyclist has to keep adjusting the front wheel, which therefore makes a wavier track than the rear wheel.

Front track

Rear track

WHEELS do two jobs. First, the rear wheel transmits the pushing force from your legs to the ground, driving the bicycle forward. Second, both wheels reduce friction by rolling over a small point of contact. The thinner the wheel and the tighter the tyre, the less the friction and the faster you go.

To steer, you just turn the handlebars the way you want to go, don't you? Wrong. Cyclists actually steer with their body weight and use the handlebars mainly for balance. In fact, to turn left you *steer right* a little first, which tips your weight to the left. This is called countersteering and it happens without thinking. Don't try doing it deliberately – you'll probably fall off!

REAR SPOKES cross each other rather than radiating straight from the hub. Being slightly off-centre helps them bear the twisting force that turns the wheel.

GEARS control the speed and force of the rear wheel. A high gear converts one turn of your feet into several turns of the rear wheel – ideal for speeding along the flat. A low gear turns the rear wheel slower but with more force – ideal for cycling uphill.

INERTIA

If you pull the front brake suddenly, your inertia can send you hurtling over the handlebars. The secret to safe stopping is to brake gradually and to keep the front and back brakes balanced. In wet weather, brake slower and longer.

TYRES provide just enough friction to grip the ground. The air inside makes them elastic, helping absorb shocks.

PEDALS convert the up-and-down motion of your legs into rotation. They also work as levers, magnifying the force from the legs to pull the chain.

How was the bicycle invented?

1817

The world's first bicycle was the "dandy horse" – a wooden running machine used as a substitute for a horse. You sat in the middle and pushed the ground with your feet, steering the front wheel by hand.

1863

Pedals were added to the front wheel in 1863, resulting in the "velocipede". You could ride this without touching the ground, but you had to pedal furiously if you wanted to go fast.

1872

For more speed, people made the front wheel bigger. The "penny-farthing" was fast but dangerously unstable. It was all too easy to fall off, and fallers tended to land headfirst. Ouch!

AIRFLOW

HANDLEBARS are levers that make it easy to turn the front wheel. The wider the handlebars, the easier it is to make fine adjustments to the wheel. Racing bikes have "drop handlebars" to help you keep your head down, which reduces drag.

DRAG

After hills and traffic, drag is every cyclist's worst enemy. It causes 70–90 per cent of the resistance you feel riding along the flat and it gets worse as you go faster. If you don't believe it, try cycling into a strong wind.

BRAKES grip the wheel rims to create friction and slow you down. The bike's energy doesn't just disappear though – it turns into sound (hence the squealing) and heat. Try touching the wheel rims after slamming on the brakes to see how hot they get.

A clever way to reduce drag is to cycle just behind another cyclist, where invisible whirlpools in the air give you an extra push. This is called "drafting" and it can cut your energy use by up to 40 per cent.

Using energy

A bicycle turns 90 per cent of the energy you supply into movement energy, making it the most efficient transport machine in the world. A car converts only 25 per cent of its energy into movement energy, and most of that is used shifting its own huge weight. In fact, the energy you use to cycle 1 km would shift a car only 20 m. You need less energy to cycle 1 km than to travel by car, train, horse, or even to run or walk.

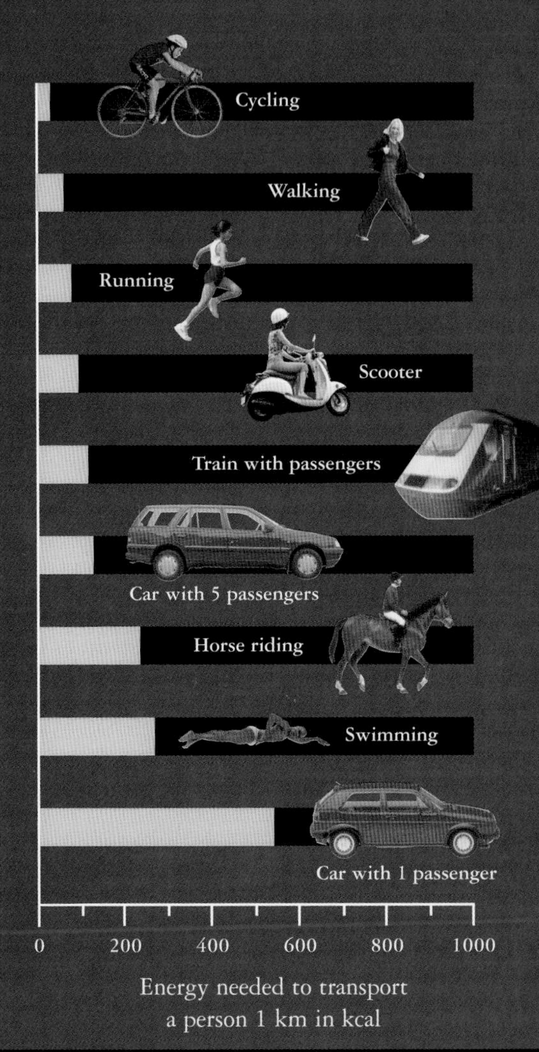

Energy needed to transport a person 1 km in kcal

1884

To make bikes safer, the front wheel was shrunk and the pedals were linked to the rear wheel by a chain and gears, making pedalling easy. The result was the "safety bicycle", the design we use today.

1893

Invented more than a century ago, the recumbent bicycle has never been as popular as the upright bicycle, despite being more comfortable, more efficient, and faster, since the low posture reduces drag.

2006

For maximum speed, a bike has to weigh as little as possible. Professional track bikes have no gears, no brakes, no handlebars, almost no tyres, and a frame made of carbon composite. As a result they weigh as little as 5 kg.

How *fast* can you FALL?

FAQ

What's your limit?

Your terminal velocity depends on your shape and weight. Something very light and fluffy, like a dandelion seed, catches so much air that it has almost zero terminal velocity and can float. The terminal velocity of rain is 27 kph (17 mph), which is about as fast as a person running. The terminal velocity of a cat is 100 kph (62 mph), which is half that of a human being and just low enough for a cat to survive a fall from the top of a skyscraper.

Can you fall faster?

Skydivers can vary their terminal velocity by changing their posture. Normally a skydiver splays his arms and legs to create as much drag as possible and to stop himself tumbling randomly. In this position, called a "stable spread", his terminal velocity is about 200 kph (124 mph) and he can open a parachute safely. To speed up, he either "stands up" or dives headfirst with his arms and legs held straight behind him. In this more streamlined posture he can reach 290 kph (180 mph).

A good way to find out how the force of GRAVITY works is to jump out of a plane. Gravity pulls your body every moment of your life, but there's usually something solid under you to counteract its effect. So what happens when there's nothing to break your fall?

Forces make things ***accelerate***. The moment you step out of plane, gravity pulls you down with as much acceleration as a *Bugatti Veyron Super Sport*, the world's fastest production sports car. In only 3 seconds you pass 100 kph (62 mph). But you don't accelerate all the way down. As you speed up, friction with the air (drag) gets stronger and stronger, until the wind roaring past feels like a hurricane. Ten seconds into the the jump, the force of drag balances the pull of gravity and you stop accelerating. You've reached **TERMINAL VELOCITY**.

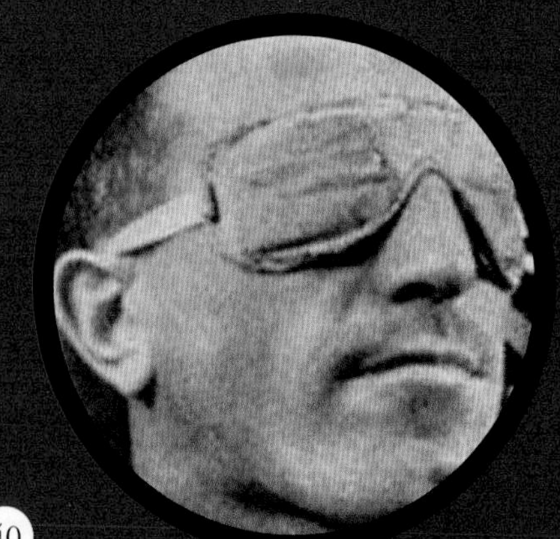

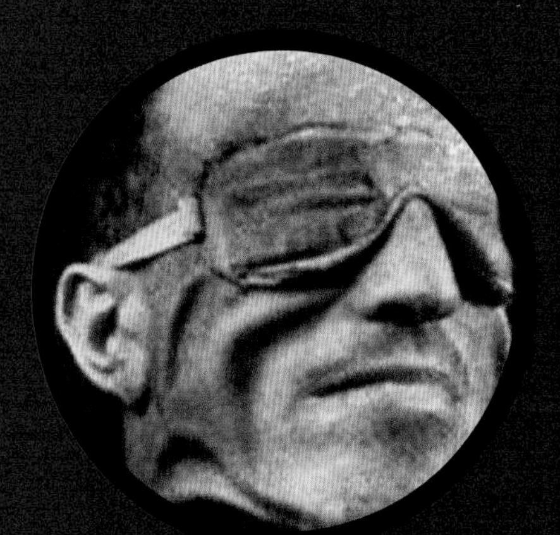

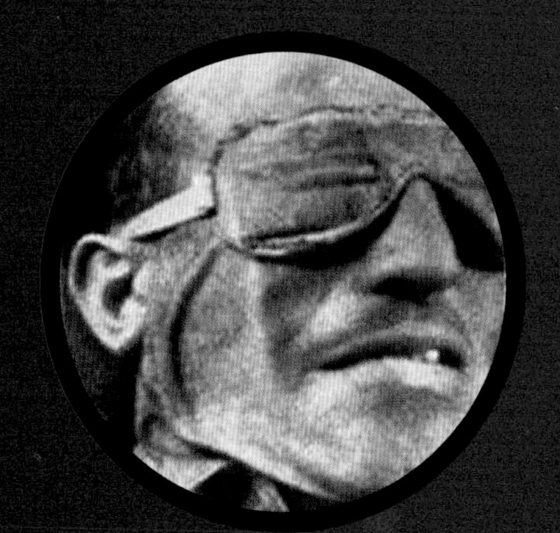

DRAG

GRAVITY

Although gravity seems like impressive stuff when you're plunging to Earth at 200 kph (124 mph), it's actually the ***weakest force in the Universe***. All objects pull on each other with gravity, but so weakly that we never notice. It takes an entire planetful of matter to create a noticeable amount of gravity.

FAQ

How do parachutes work?

Parachutes work by increasing the force of drag. When a parachute first opens, the sudden increase in drag makes the skydiver decelerate violently until his velocity drops to about 20 kph (12 mph). If there were no air, a skydiver would accelerate all the way down, even with his chute open, and hit the ground at more than 1000 kph (620 mph).

What's the highest skydive?

The world's highest skydive took place on 24 October 2014 when Google executive Alan Eustace leapt from a height of 39,625 m (130,000 ft), wearing a specially designed spacesuit. Eustace reached a top speed of 1300 kph (807 mph), breaking the sound barrier. Air is thinner at great heights and causes much less drag.

Blowing in *the wind*

This is what the force of drag can do to a human face without the protection of a mask or visor. As the wind speed increases from 440 kph (275 mph) to 560 kph (350 mph), the blasting air stretches your skin and rips your mouth open. Skydivers never experience this much drag, but fighter pilots can do if they eject from jet aircraft in an emergency.

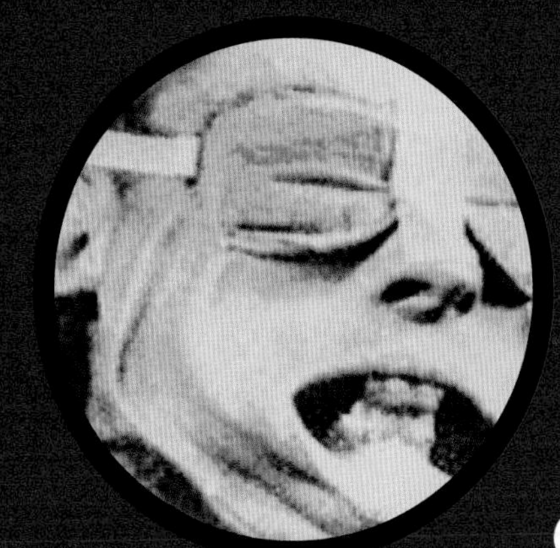

How do PLANES

FAQ

What's the best shape for flight?

The ideal shape for a plane depends on what job it has to do. Long wings make planes more efficient but slower and harder to turn. Short wings are better for manoeuvrability.

Fighter jets like the Eurofighter Typhoon have short, stubby wings that allow them to make very sharp turns in the air.

The bigger the cargo hold, the more a plane can carry. The Boeing Super Guppy can swallow and carry whole planes up to 26 tonnes in weight.

Since their wings rotate, helicopters don't have to move forwards to create lift. As a result, they can hover and fly backwards or sideways.

Flapping wings make insects more agile than any aircraft. Bees flap their wings 200 times a second but some midges flap 1000 times a second.

If you flatten your hand, angle it up a little, and very carefully hold it outside the window of a moving car, you'll feel a force pushing it upwards. The force is ***LIFT*** and it keeps planes in the air. So where does the force of lift come from?

Very long wings create lots of lift but little drag, allowing a glider to stay airborne without an engine to push it forwards.

THE WING of a plane isn't just angled upwards like your hand was – it also has a special shape called an aerofoil, which is rounded on top but flatter on the bottom. Both the angle and the shape cause air to flow faster over the top of the wing than beneath it. Fast-moving air has lower pressure than slow air, so the wing feels a push from below. This push is the force of **LIFT**.

stay in the AIR?

Newton's third law tells you where lift comes from. When you hold your hand out of the car window, it pushes the flowing air down. So the air must be pushing back up with an equal and opposite force. Your hand is also pushing forward, and that's matched by an equal and opposite force pushing it back – ***DRAG***.

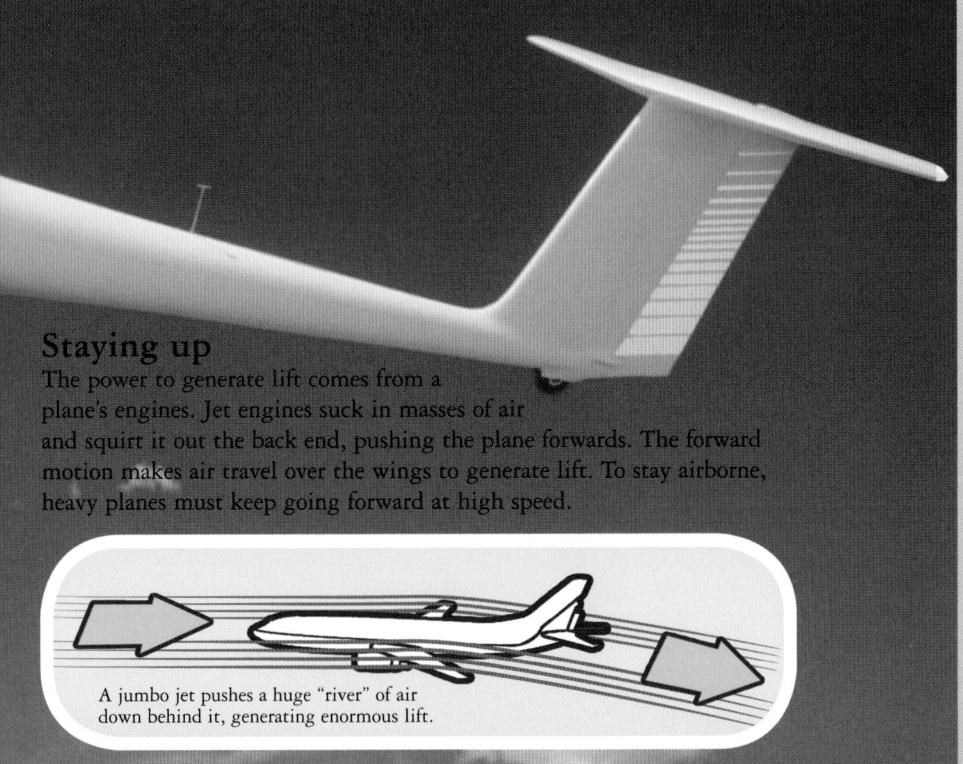

Staying up

The power to generate lift comes from a plane's engines. Jet engines suck in masses of air and squirt it out the back end, pushing the plane forwards. The forward motion makes air travel over the wings to generate lift. To stay airborne, heavy planes must keep going forward at high speed.

A jumbo jet pushes a huge "river" of air down behind it, generating enormous lift.

NOW TRY THIS!

Levitating ball trick

Planes rely on the fact that slow-moving air has a higher pressure than fast-moving air. This is called the Bernoulli effect, and you can see it for yourself by performing a magic trick. Place a ping-pong ball in the flow of air above a hairdryer and the ball will get trapped. When it tries to fall out, the slow-moving air outside the main flow pushes it back because it has greater pressure.

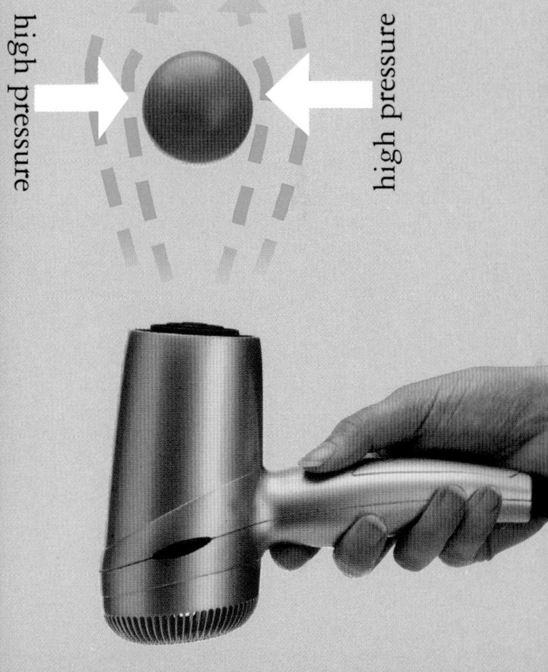

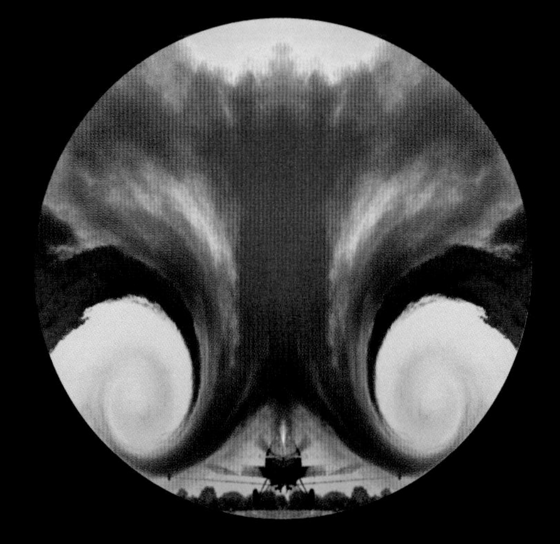

The *vortex* behind a *jumbo jet* can flip a small plane UPSIDE DOWN

What a waste

At the tip of a plane's wing, high-pressure air from below the wing swirls up to the low-pressure area on top. The swirling movement creates a whirlpool in the air – a vortex. The vortex wastes the plane's energy and so increases the force of drag.

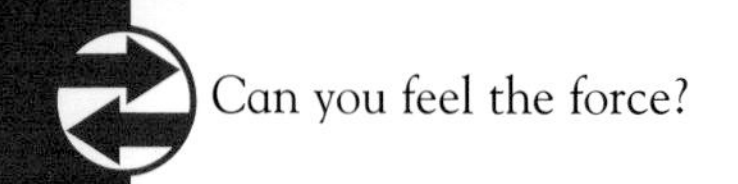

Why do **GOLF BALLS** have dimples?

While planes usually need to be as smooth and streamlined as possible to fly, the opposite is true of golf balls. The 300 or so dimples on a golf ball make it fly up to three times farther than a smooth ball would. Here's how...

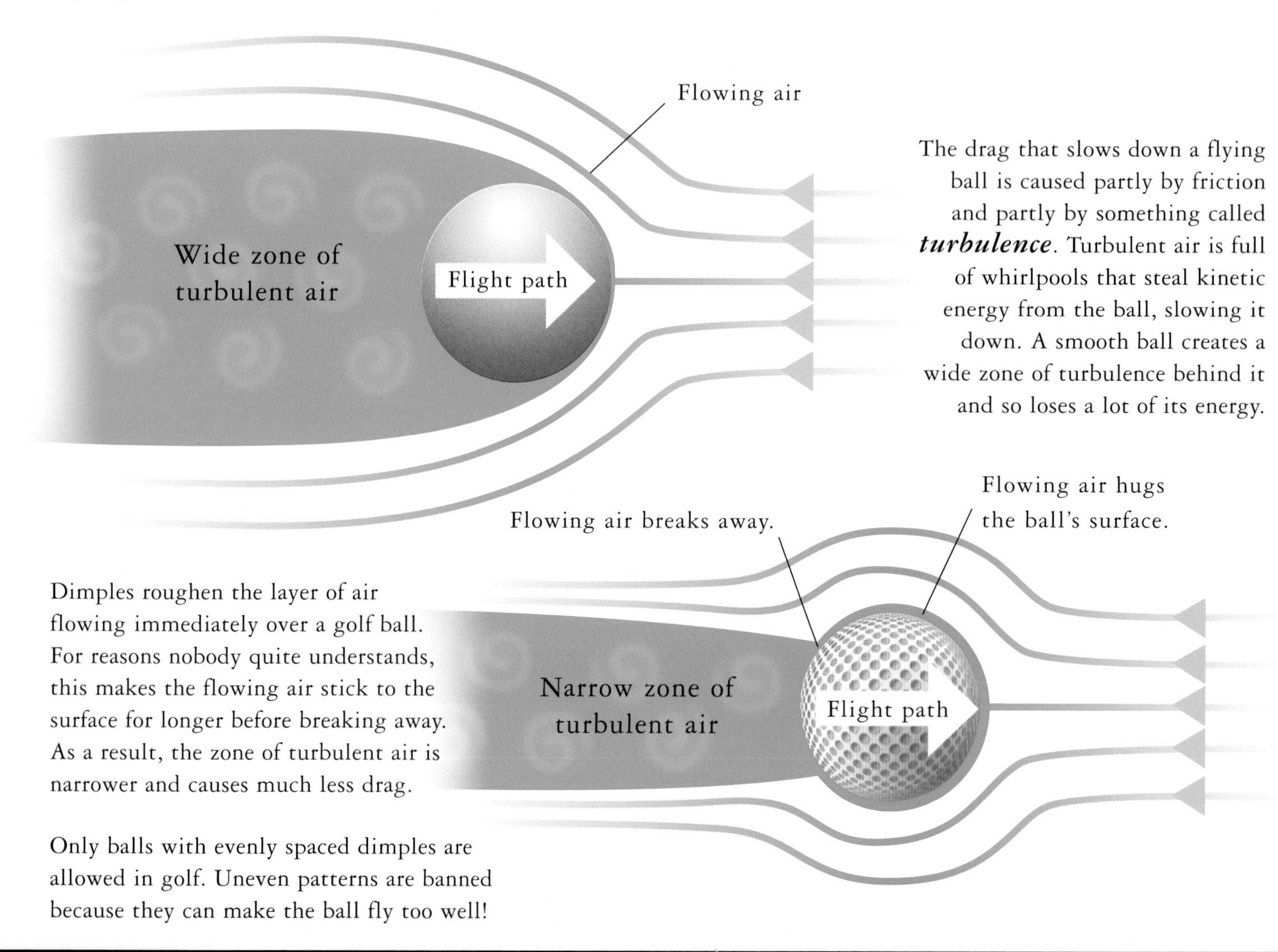

The drag that slows down a flying ball is caused partly by friction and partly by something called ***turbulence***. Turbulent air is full of whirlpools that steal kinetic energy from the ball, slowing it down. A smooth ball creates a wide zone of turbulence behind it and so loses a lot of its energy.

Dimples roughen the layer of air flowing immediately over a golf ball. For reasons nobody quite understands, this makes the flowing air stick to the surface for longer before breaking away. As a result, the zone of turbulent air is narrower and causes much less drag.

Only balls with evenly spaced dimples are allowed in golf. Uneven patterns are banned because they can make the ball fly too well!

How do bullets work?

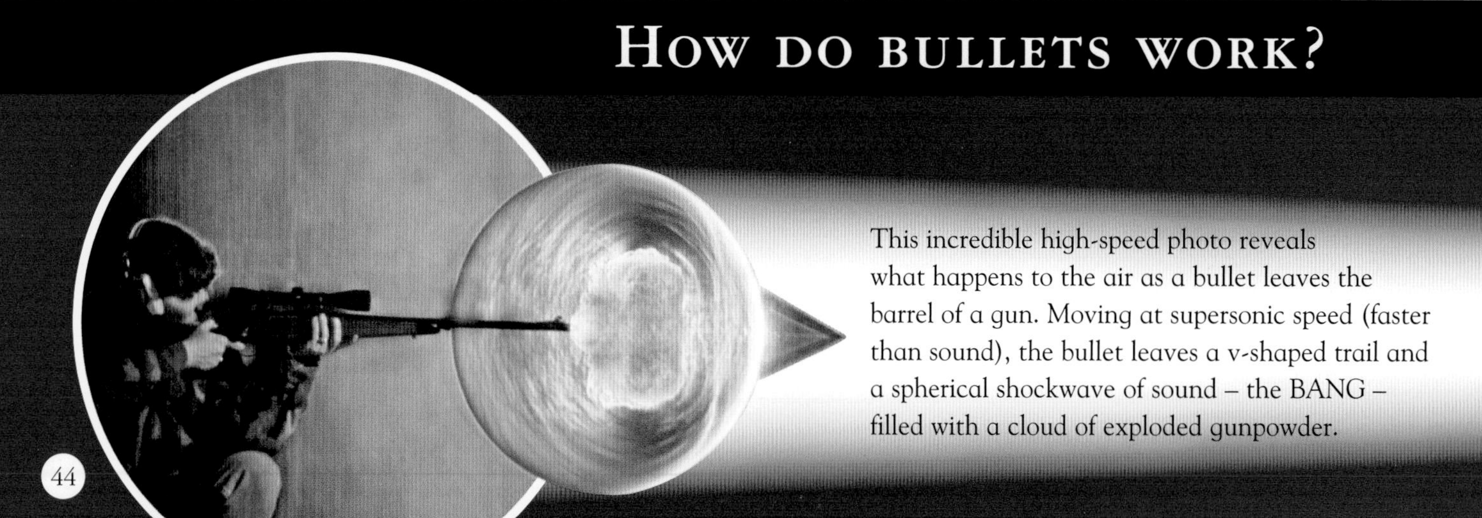

This incredible high-speed photo reveals what happens to the air as a bullet leaves the barrel of a gun. Moving at supersonic speed (faster than sound), the bullet leaves a v-shaped trail and a spherical shockwave of sound – the BANG – filled with a cloud of exploded gunpowder.

Can you *bend it like* BECKHAM?

A professional footballer can not only make a football curve but can kick it so it seems to be going straight but turns at the last moment and slips inside the goalpost. The secret is a trick called the MAGNUS EFFECT.

5. Because the air pressure is higher on the right side, the ball is pushed to the left and bends as it flies.

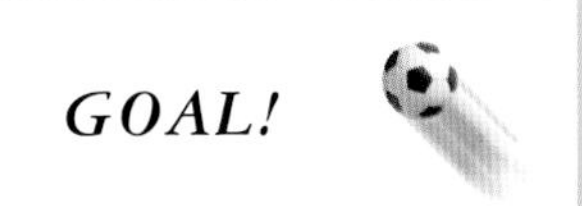

3. On this side the whirlpool of air moves *against* the oncoming air, which increases the air pressure.

4. On this side the whirlpool moves the *same way* as the oncoming air, which reduces the pressure.

2. The spinning ball drags the layer of air on its surface around with it, forming a whirlpool of air.

1. The footballer strikes the ball off centre to make it spin.

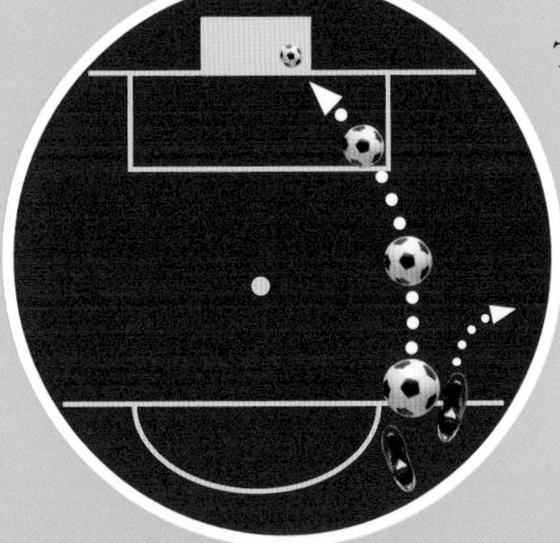

The Magnus effect works best at certain speeds. A very fast ball hardly curves at all, but a slow ball curves strongly. A skilled footballer can control when the curve will happen by giving the ball just the right combination of speed and spin. As a result, he can make a ball fly almost straight and curve into the net at the last moment, as it loses speed.

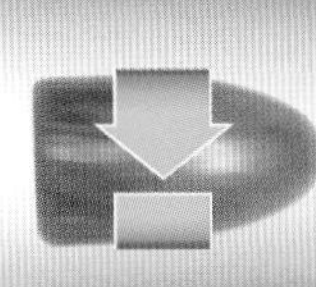

A gun's barrel has a spiral groove cut into it. The groove makes the bullet spin as it flies, creating a special kind of inertia (gyroscopic inertia) that stops the bullet tilting. A spinning bullet follows a very straight path, giving it deadly accuracy.

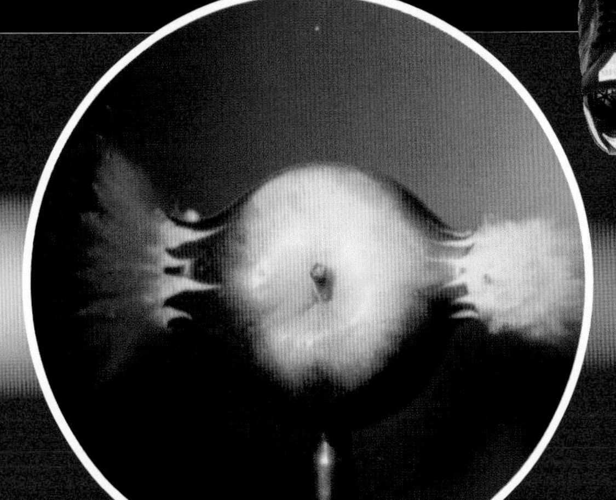

Though tiny, bullets have phenomenal kinetic energy due to their speed, which can easily exceed 1000 kph (620 mph). It's this energy that makes a bullet so destructive.

What's the best *shape*

The shape of cars has changed a lot in the last 100 years. The first cars looked like giant prams, sounded like tractors, and were barely faster than walking speed.

FAQ

Why don't cars take off?

For fast cars, staying on the ground can be a challenge. Air tries to flow under as well as over a car, which can make the whole vehicle act like a wing and try to lift off as speed rises. A 360-tonne jumbo jet only has to reach 290 kph (180 mph) to take off, so how does a 1-tonne Ferrari stay down at 300 kph (186 mph)? One solution is to add a spoiler – an upside-down wing on the rear of a car. A spoiler creates negative lift, pushing the rear wheels down and improving grip at the same time.

Speed machines

As with planes, cars are shaped to suit the job they do. Dragsters are built purely for acceleration in a straight line. A very long body and spoilers at both the front and rear counter the dragster's tendency to lift off the road. Massive rear wheels deliver the pushing force, accelerating the car from 0 to 530 kph (330 mph) in under 5 seconds.

Dragster

Keep a low profile

Aerodynamics is the study of the way air flows around objects. A car has to push a hole through the air as it moves, which gets increasingly difficult as speed increases. So the first rule of aerodynamics is to keep the hole small. Instead of making a car tall and square, make it long and flat.

Wind tunnels

In the past, designers built clay or metal models of cars and tested their aerodynamics in a wind tunnel (left), which blows a jet of air and smoke over the car to reveal the shape of the airflow. These days the job can be done by computers that model the physics of flowing air (right), allowing a car's shape to be perfected without the expense of a real model.

for a CAR?

Today's supercars are altogether sleeker, flatter, curvier, and of course faster. So why do they look so different? In a word: AERODYNAMICS.

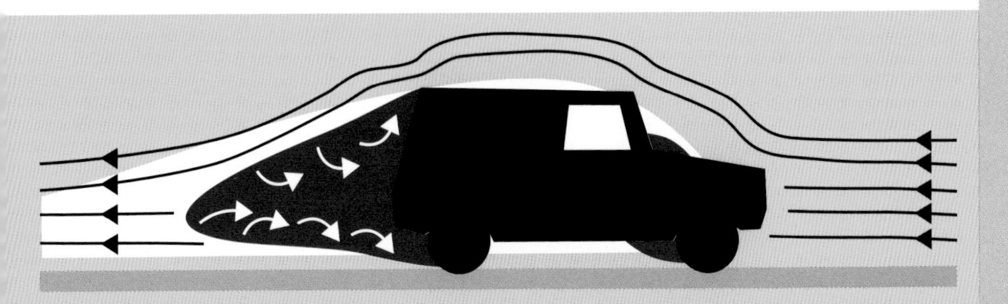

DON'T STIR THINGS UP

The next rule is to help the air flow as smoothly as possible. Like golf balls and planes, cars leave a trail of swirling, ***turbulent*** air behind them, which wastes energy. Boxy shapes with sharp corners create lots of turbulence; smooth, tapering shapes create less.

A Formula One car gets a workout in a computer-generated airflow test. The program can isolate different parts of the airflow, such as the air moving through the engine bay and around the wheels as they spin.

DESIGN TIPS

Keep the weight down

The lighter a car is, the faster it accelerates. The Ariel Atom is a third the weight of an average car and can reach 100 kph (60 mph) in 2.9 seconds. To keep the weight down it has no roof or body (but there is plenty of room for an umbrella).

Lie flat

Solar-powered cars are as flat as pancakes to minimize drag, but the driver has to lie down. The flat shape also provides room for the car's solar panels.

Channel the airflow

By taking in air at the front and channelling it at speed under its bodywork, the Pagani Zonda creates a low pressure zone below that helps keep its lightweight carbon body pressed down.

What not to drive

Before car designers started using wind tunnels, they made cars *look* streamlined but got the physics wrong. The Lamborghini Miura of the 1960s was the top supercar of its day. It was sleek, sophisticated, and fast – but totally unsteerable above 240 kph (150 mph) because the wheels wouldn't stay on the road.

Why do balls bounce?

If you drop a beanbag on the ground, it loses all its kinetic energy and sits there in a heap. But a bouncy ball squeezes like a spring and stores its energy as ***potential energy***. As it rebounds, the stored energy turns back into kinetic energy and pushes the ball up again.

warm balls bounce highest

A tennis ball loses nearly half its kinetic energy each time it hits the ground, so each bounce is only about half as high as the last one.

ELASTIC

Newton's cradle

Objects that conserve kinetic energy when they hit something are called ***elastic***. Steel balls are so elastic they can pass their energy along a row with very little loss. In a "Newton's cradle", when one ball is swung to hit the others, its energy passes along the line and the last ball springs outward as if by magic.

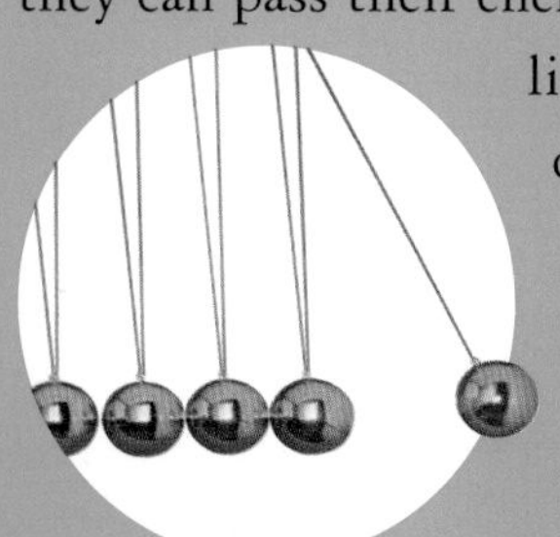

Try the two-ball bounce

Take two bouncy balls of different sizes and put one on top of the other. Drop them on a hard surface and watch what happens. The bottom ball hits the ground first and then transfers its kinetic energy to the top ball, making it bounce much higher than it would on its own.

When a ball bounces off the ground, *planet* ***Earth*** bounces the opposite way a tiny bit

BOUNCE-ability

Balls bounce best when they store energy by compressing rather than bending. Something bendy, like a beanbag, won't bounce at all. Solid balls bounce well (even marbles), and air-filled balls bounce well only when properly inflated. The bounciest ball of all is a steel ball, like the one in Newton's cradle. It can bounce to 98% of its original height if you drop it on solid steel.

0% BEANBAG

30% BASEBALL

40% FOOTBALL

56% TENNIS BALL

56% BASKETBALL

67% GOLF BALL

81% SUPERBALL

98% STEEL BALL

INELASTIC

Why don't dogs bounce?

Objects that don't conserve kinetic energy when they hit something are called ***inelastic***. Dogs and people are inelastic because they don't immediately spring back into their former shape on hitting something. But they will bounce if they hit a very elastic surface, such as a trampoline.

Inelastic collisions

When two inelastic objects hit each other at high speed, the kinetic energy goes into changing their shapes rather than making them bounce apart. In a car crash this is actually a good thing. By crumpling up, the cars absorb most of the energy and so protect the people inside.

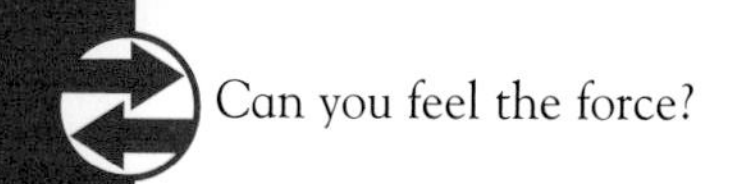

Can you *lie* on a bed of nails?

A bed of nails is made of thousands of nails pointing upwards. Surely you'd get STABBED thousands of times if you tried to lie down on one? Actually you wouldn't, provided you got on the bed carefully. To find out why not, you have to understand about PRESSURE.

Under pressure

The amount a pushing force is concentrated or spread out is called *pressure*. When you push a drawing pin into a wall, the force from your finger is concentrated into a far smaller area at the other end. As a result, the pin goes into the wall and not your finger. The sharper the point of the pin, the greater the pressure, because:

pressure = force ÷ area

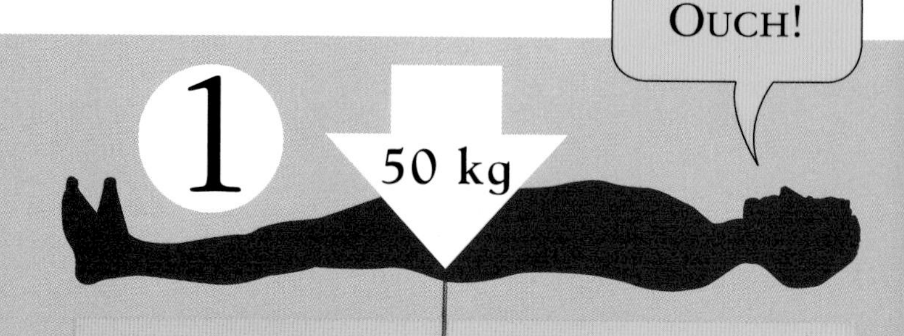

Now think about how pressure works with a bed of nails. Let's suppose you weigh 50 kg and you lay on a "bed of one nail". The force of all your weight would be concentrated on one nail. It would hurt – a lot.

Now suppose you lay on a bed of 5000 nails and your body touched about half of them. The weight on each nail would be 50 kg ÷ 2500, which is [illegible] It wouldn't hurt at all.

Now try this

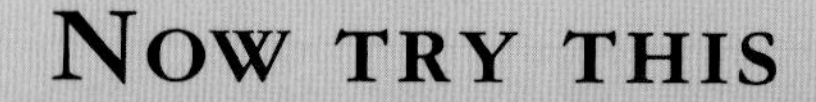

Shrink-wrap your arm

To witness the awesome power of air pressure, remove the sleeve from one of your arms and place your arm in a dustbin liner. Use the suction tube of a vacuum cleaner to suck out the air. The pressure of the air outside will press the plastic tight

Make a bed of nails

You can see how a bed of nails works with a packet of drawing pins and a tomato or a toy animal. (Don't use a real animal.) Start with one pin and see what happens when you let the tomato's weight rest on it. Increase the number of pins to reduce the pressure until the tomato's weight

Water pressure

Even liquids and gases put pressure on things. When you swim underwater, water molecules press against your body. As you go deeper, the weight of the water above you makes the pressure rise. At 1 km (0.6 miles) depth, the force on every square centimetre of your skin is about 1 tonne (7 tons per square inch). Deep-sea divers need special diving suits to withstand such fantastic pressures.

To see how water pressure rises with depth, cut the top off a plastic bottle, make a series of holes down the side, and fill it with water. The bottom jet squirts farthest because pressure is highest.

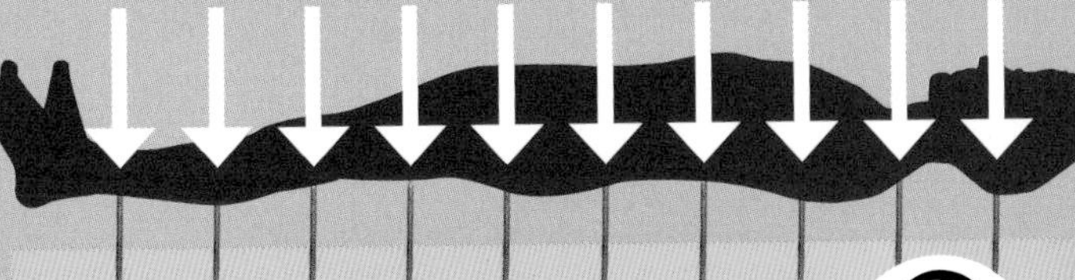

2

If you lay on a "bed of ten nails", the pressure on each nail would be a tenth as much, but still a lot. It would be about the same as putting a watermelon on each nail.

3

Air pressure

As you read this, the air around you is pushing on your body with a force of about 15 tonnes. If the atoms in your body weren't pushing back equally, you'd be crushed in an instant. When the pressure of air goes up and the volume stays the same, the temperature automatically rises. As you pump up a bike wheel, you can feel the heat of the pressurized air coming through the pump.

Secret squirter

Here's a practical joke to play on nosy parkers. Write "DO NOT OPEN" with a marker pen on a plastic bottle and make holes in the letters with a pin. Fill with water and screw on the lid. The holes won't leak because air pressure keeps the water in. But if a nosy parker secretly opens the bottle, air gets in the top and they get soaked!

Do NOT OPEN

And for my next trick...

Fill a glass tumbler to the brim with water and slide a postcard over the top. Keeping your hand pressed on the card, turn the tumbler upside down over the sink. Take your hand away and the card *should* stay in place, held in place by air pressure!

What's the MATTER?

Imagine you chopped an apple in half, chopped the halves in half again, and then carried on doing the same millions and millions of times. Let's assume you had a pretty small knife.

After a few billion years of chopping you'd get to a point where you can't chop any further. You'd have reached the building block that makes up everything in the Universe: THE MIGHTY ATOM.

For years, scientists thought atoms were the smallest things possible.

Then someone found you can chop even further, and a strange new world opened up...

What is MATTER

EVERYTHING is made of atoms. Houses, trees, cars, dogs, your breath, your body, the rain, the air are ALL MADE OF ATOMS.

The average ATOM can last a hundred

How big is an atom?

Atoms are pretty small. Half a million lined up in a row could hide behind a hair, and it takes 300 billion billion atoms to make one drop of water. To see the atoms in the water drop below, you'd need to enlarge the photo to 320 km (200 miles) wide.

What do atoms look like?

An atom doesn't look like anything. It's far too small to reflect light, so the question makes no sense. Even so, it is possible to take pictures of the electric field around single atoms. The picture below shows a heap of gold atoms (red and yellow) on a layer of carbon atoms (green).

How long do atoms last?

Atoms are almost indestructible. When you die, the atoms in your body don't die with you – they get recycled. About a billion of your atoms were once in the bodies of Julius Caesar, Jesus Christ, and Aristotle. Nobody knows for sure how long atoms can last, but one eminent scientist reckons that 100,000,000,000,000,000,000,000,000,000,000,000 years is a sensible estimate.

You've got a billion bits of me in you!

Aristotle

made of?

Nearly every atom in your BODY was part of millions of other creatures before it got to YOU.

million billion billion billion years

How were atoms discovered?

The first really good evidence for atoms came from the discovery that certain chemicals always combine in particular proportions (because the atoms join in ratios to make molecules). Later, scientists found they could explain the way pressure and temperature change in a gas if you assume the gas is made of billions of tiny bouncy particles.

Can you split atoms?

When scientists discovered atoms about 200 years ago, they thought atoms were the smallest bits of matter possible and therefore impossible to split. But atoms do split – just rub your hair and you'll split some. Splitting the middle bit of an atom (the nucleus) is very difficult though. And it can cause a nuclear explosion.

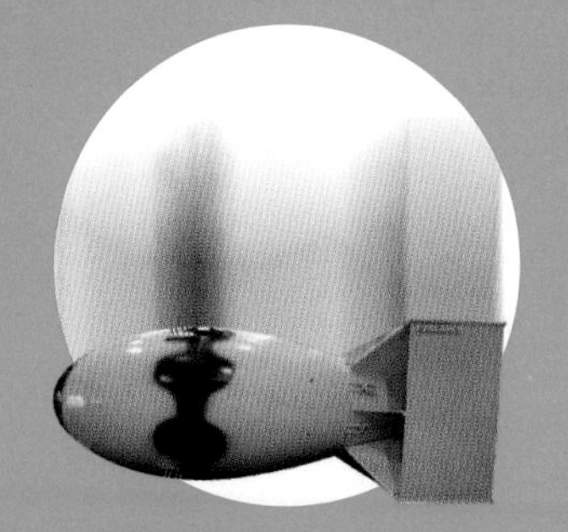

What are molecules?

Atoms tend to stick together. They pull on each other with powerful forces and glue themselves into clumps called molecules. Water molecules (H_2O) are made of three atoms each – one large oxygen atom glued to two small hydrogen atoms. Nearly everything we see and touch is made of molecules.

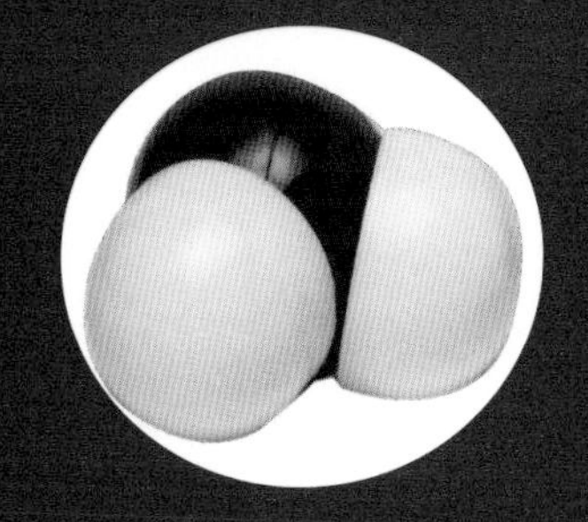

What's *inside* an ATOM?

Atoms are a bit like Russian dolls. Scientists used to think they were the smallest bits of matter possible, but then people started finding all sorts of smaller bits inside them. Inside the nucleus are particles called protons and neutrons, and inside them are quarks. That may be far as it goes, but some physicists think there could be even deeper levels still.

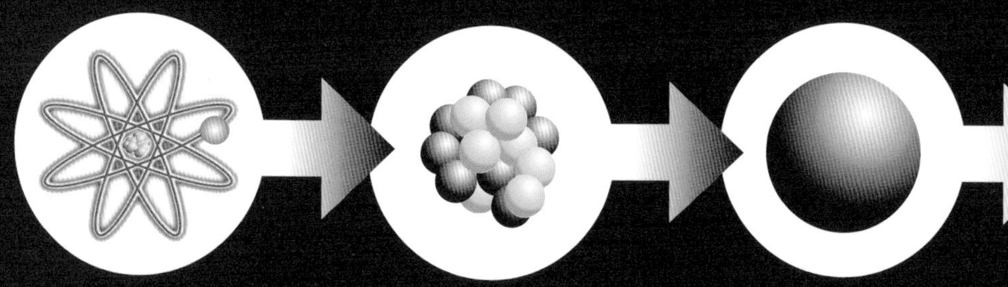

Deep inside an atom is a ZOO of even smaller particles. They inhabit a weird world where the normal laws of physics don't work.

Electron

Electrons whizz around the nucleus, trapped by a pulling force from it. People once thought electrons orbited like planets round the Sun, but the truth is weirder. An electron is never really in one place at a time. It exists in lots of places at once as a sort of cloud of possibility. Weirder still, an electron can leap to a new orbit without passing through the space between – a "quantum leap".

Neutron

Neutrons are like protons but have no charge. In fact, a neutron can turn into a proton by spitting out an electron, which leaves it positive. Most atoms have about as many neutrons as protons, but big atoms have an excess of neutrons. The extra neutrons increase the strong nuclear force to hold all the protons together. If taken out of an atom, a neutron lasts 886 seconds and then falls apart.

Nucleus

The nucleus is the solid centre of the atom, where nearly all the atom's weight is concentrated. It's surprisingly tiny, taking up only one millionth of a billionth of the space in the atom. If the whole atom were as big as a cathedral, the nucleus would be the size of a fly. Since electrons are even smaller (and in fact have zero size), an atom consists almost entirely of empty space.

Quark

Each proton and neutron is made up of three even stranger particles called quarks. Quarks can't exist on their own – they only come in pairs or triplets. Sometimes, pairs of quarks emerge out of nothingness for no reason whatsoever. The strong nuclear force that holds the nucleus together is transmitted by pairs of quarks emerging from nothing and flying about.

Proton

The nucleus is made of two types of particle: protons and neutrons. Protons have a positive charge that pulls on the electrons, which are negative. The nucleus can hold more than 100 protons, so why don't all these positive particles repel each other and blow the atom apart? The answer is that they're held together by a powerful force that only exists inside atomic nuclei: the strong nuclear force.

String

Imagine you shrank to the size of an atom, shrank by the same amount again, and then shrank by the same amount a third time. What would you see? According to the latest big theory, you'd see loops of "string". String theory says that all the particles in atoms are vibrations on the same kind of string, just as a violin string can create all the different musical notes.

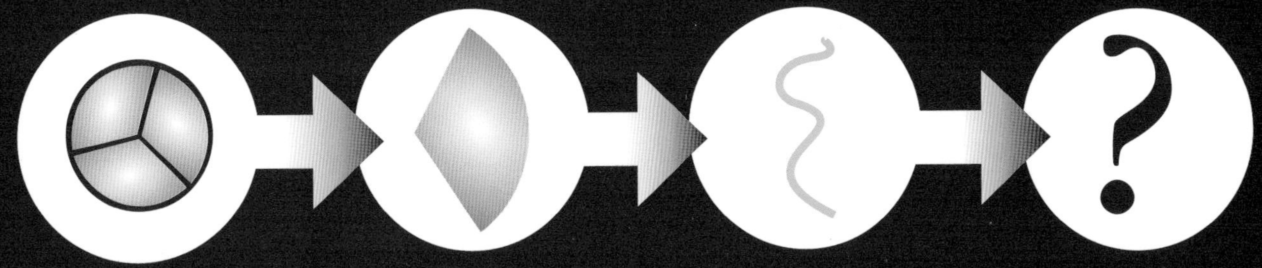

Why do *balloons* STICK to the wall?

FAQ

What happens when you switch on the light?

While balloons are good at trapping electrons, other substances – such as metals – let electrons flow through them. This is called current electricity. When you switch on a light, you connect two wires together and so let current electricity flow around a circuit. The moving electrons carry energy and power the bulb.

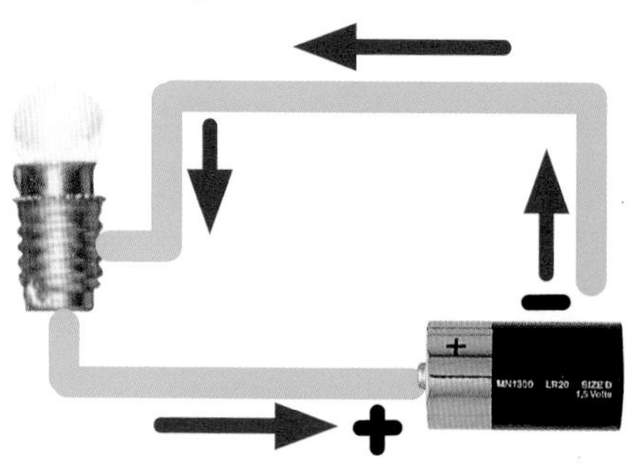

How fast do electrons flow?

People once thought electricity is like flowing water, but in reality it's more like a row of billiard balls. If you hit the ball at the end, the force is passed along in a relay. In a wire, electrons nudge each other in a similar way. The electrons themselves move slower than a snail, but their energy zips through at the *speed of light*.

WHAT IS ELECTRICITY?

Electrons normally stay in atoms because they are held there by a FORCE. Electrons have a negative charge and the nuclei of atoms have a positive charge. Opposite charges pull on each other (a bit like opposite poles of a magnet), and like charges push each other away (they repel). Because electrons and nuclei have opposite charges, they pull on each other and stick together, forming atoms. But they don't always stay together. Sometimes electrons can break off atoms and carry their charge elsewhere, and that's what causes electricity.

Can you make your hair stand on end?

We normally think of electricity flowing through wires, but sometimes electrons get stuck in one place. When that happens, it causes something called ***static electricity***. It's easy to create static electricity by rubbing objects together. If you rub your hair with a polyester t-shirt, the t-shirt will strip electrons from your hair. Your hairs become positively charged and repel each other, which makes them stand on end.

A brief *history* of ELECTRICITY

The ancient Greeks had discovered static electricity in amber by 600 BCE.

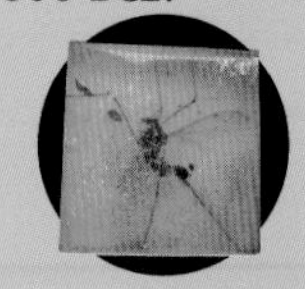

1700

Around 1700, scientists discovered how to build machines that create powerful sparks of static electricity.

1730

English scientist Stephen Gray suspended a boy from a silk rope and charged him with static, proving that the human body can be charged like amber.

1752

American statesman Benjamin Franklin flew a kite into a storm cloud to prove that lightning is caused by electricity. He coined the terms positive and negative charge (and got them the wrong way round.)

Rub a balloon on your hair and hold it to the wall. It will stick there, but what's holding it? It's the same thing that traps electrons around the nuclei of atoms: ELECTRICITY.

How do balloons stick to the wall?

Static electricity is what makes balloons stick to the wall after rubbing them on your hair. As you rub a balloon, electrons break off your hair and stick to the rubber, giving it a negative charge. When you press the balloon to the wall, the electrons push away electrons in the wall and make its surface positive. And so the negative balloon sticks to the positive wall surface.

Opposite charges *attract*, similar charges *repel*

Why are carpets shocking?

The electrons trapped in static electricity will try to escape as soon as they touch something that lets them flow away. This can happen so quickly that it causes a spark or an electric shock. If you have shoes with plastic soles and walk on a nylon carpet, your body picks up electrons. When you touch something metal like a door handle, the charge rushes through you and gives you a shock.

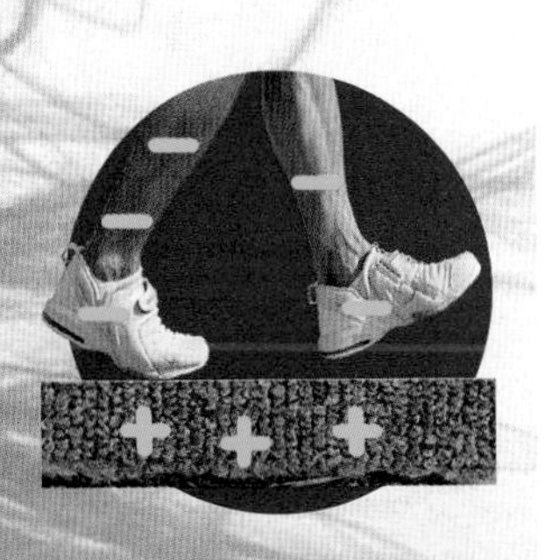

FAQ

How does lightning work?

Lightning is caused by static electricity. Inside a storm cloud, ice crystals race up and down and rub each other. For unknown reasons, this makes electrons collect at the bottom of the cloud, making it negative. The charge is so vast that it forces its way to the ground by tearing the electrons off air atoms to create a charged path – a bolt of lightning.

How wide is a bolt of lightning?

A bolt of lightning is only 2–3 cm (1 in) wide but carries a huge amount of energy. Travelling at 435,000 kph (270,000 mph), it heats the air to 28,000°C and makes it explode, causing the bang we hear as thunder. Most bolts carry negative charge, but occasional positive bolts strike from the cloud tops. They are far deadlier and can be many kilometres in length.

1753

Russian scientist George Richmann tried to repeat Franklin's experiment and was struck by lightning and killed.

1771

Italian scientist Luigi Galvani discovered that frogs' legs twitch when given an electric shock, and concluded that electricity is the essence of life. The theory inspired the story of Frankenstein.

1800

Alessandro Volta invented the battery after investigating Galvani's twitching frogs' legs.

1879

After thousands of experiments, Thomas Edison perfected the light bulb.

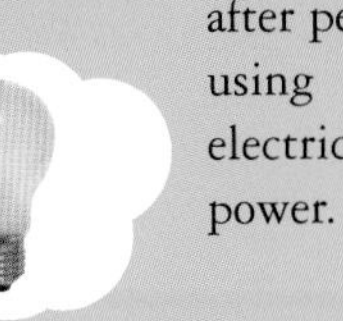

1897

English scientist J.J. Thomson discovered the electron in 1897 – years after people had started using electric power.

Shocking experiments

Leaping men

Draw a tiny outline of a man on thin paper and cut him out through several sheets of paper to make a group. Scatter the men on a table. Rub a balloon or CD box quickly on your hair for 30 seconds and then lower it over the men.

What's going on?

The balloon picks up electrons from your hair and becomes negative. As you lower it, the negative charge pushes away electrons in the paper men and makes their tops positive. Opposites attract, so the men leap up to the balloon.

Electric flea circus

Scatter mustard seeds, rice, or paper circles from a hole punch on a sheet of paper. Rub the lid of a CD box on your hair for 30 seconds. Hold the lid a finger's height over the seeds and lower it very slowly. The seeds will dance up and down like jumping fleas.

What's going on?

The CD lid becomes negatively charged and pushes electrons in the seeds towards the bottom, making their tops positive. They jump up and stick to the lid, but their charge then seeps away and they drop off. The process repeats, making them dance up and down.

Bending water

Turn on a cold tap and slowly close it to make a smooth, thin stream of water. Rub a balloon or a plastic object on your hair to charge it with static. Hold the charged object near the water and watch what happens.

What's going on?

The balloon or plastic picks up electrons and becomes negative. When you hold it near the water, it repels electrons on the near side of the water, making that side of the water positive. The positive side is drawn towards the balloon, making the stream of water bend.

POSITIVE THIS END

Some materials are good at giving away electrons and becoming positive when you rub them, but others are better at taking electrons and becoming negative.

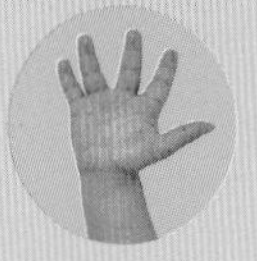

Skin

Rabbit fur

Glass

Hair

Nylon

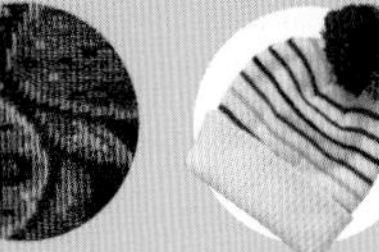

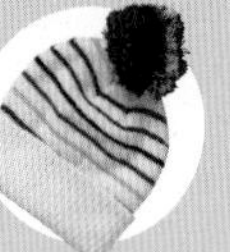

Wool

Cat fur

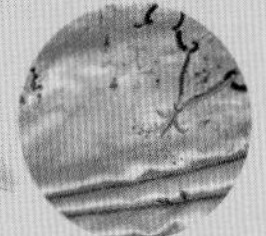

Silk

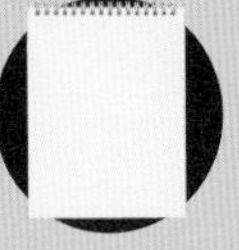

Paper

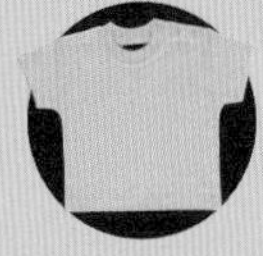

Cotton (neutral)

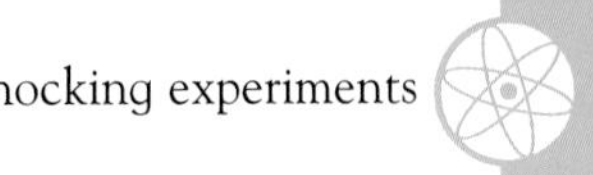

Shock your family, friends, and yourself with these ELECTRIFYING experiments! Most of them work best when the air is very dry – a sunny winter's day is perfect. In humid or wet weather, moisture in the air makes static electricity seep away. Tip: make sure you hair is clean and dry.

Sour power

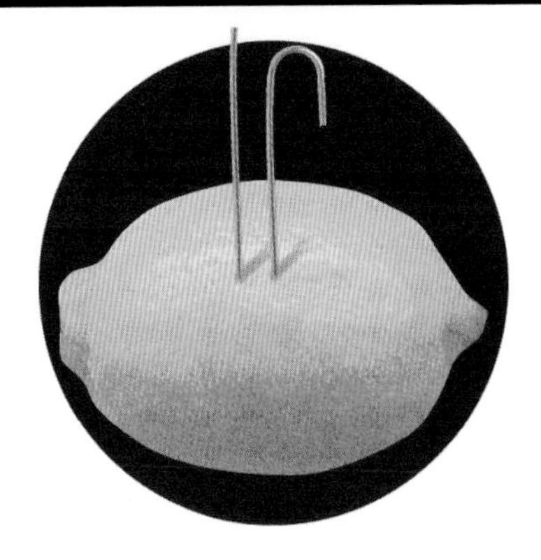

Gently squeeze a lemon, then stab it with a straightened paperclip and an equal length of copper wire, placed close together. (Ask an adult to strip an old cable to make the copper wire.) Press your tongue to both at once and you'll feel a tingling sensation.

What's going on?
The lemon is a battery. Acid inside it reacts chemically with both metals. Electrons leak away from the copper (making it positive) and build up on the paperclip (making it negative). When you lick the wires you complete the circuit and the charge flows across your tongue.

Snake charmer

Draw a spiral pattern on a piece of lightweight paper and cut along the line. Rub a plastic pen vigorously on your hair or your jumper for 30 seconds. Hold the pen near the middle of the spiral and slowly lift it.

What's going on?
The pen picks up electrons from your hair and becomes negatively charged. When you hold it to the paper, it repels electrons. The paper then becomes positive and sticks to the pen, rising as you lift it – just like a snake being charmed!

Sparks in the dark

1. Fully extend the aerial on a radio (not a digital one). Charge a balloon on your jumper and check that it sticks to your jumper. Now slowly bring it towards the aerial and listen. What do you hear? Does the balloon still stick to your jumper?
2. Switch on the radio, set it to AM or medium wave and tune it away from a station. Turn the volume to full. Charge the balloon and bring it slowly to the aerial again. What do you hear?
3. With the room in total darkness, charge the balloon and touch the aerial one more time. What do you see?

ANSWERS
1. You hear a crack as the electrons leap to the aerial. The balloon no longer sticks to your jumper because the charge has gone.
2. You hear a much louder pop because the radio picks up the spark and turns it into sound.
3. You see tiny sparks as the charge leaps through the air – lightning in miniature.

When arranged in order, these materials make what scientists call the "triboelectric series". To generate a good static electric charge, rub things from opposite ends of the series together.

NEGATIVE THIS END

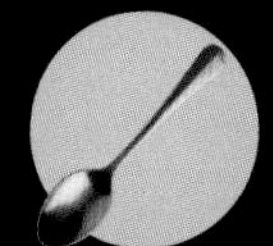
Steel (neutral)

Wood

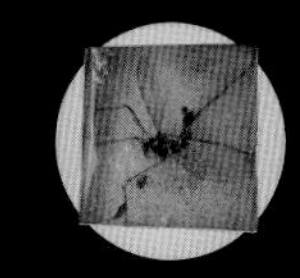
Amber

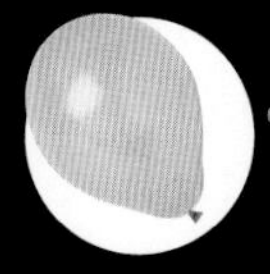
Rubber

Brass

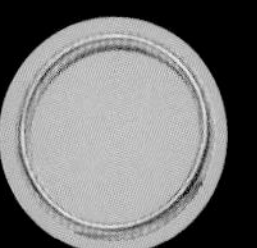
Gold

Polyester

Polystyrene

Polythene

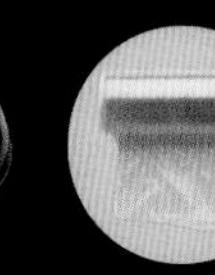
Clingfilm (PVC)

How do *MAGNETS*

The electrons that whizz around ATOMS don't just create electricity. They are also responsible for the ***mysterious** force* of magnetism.

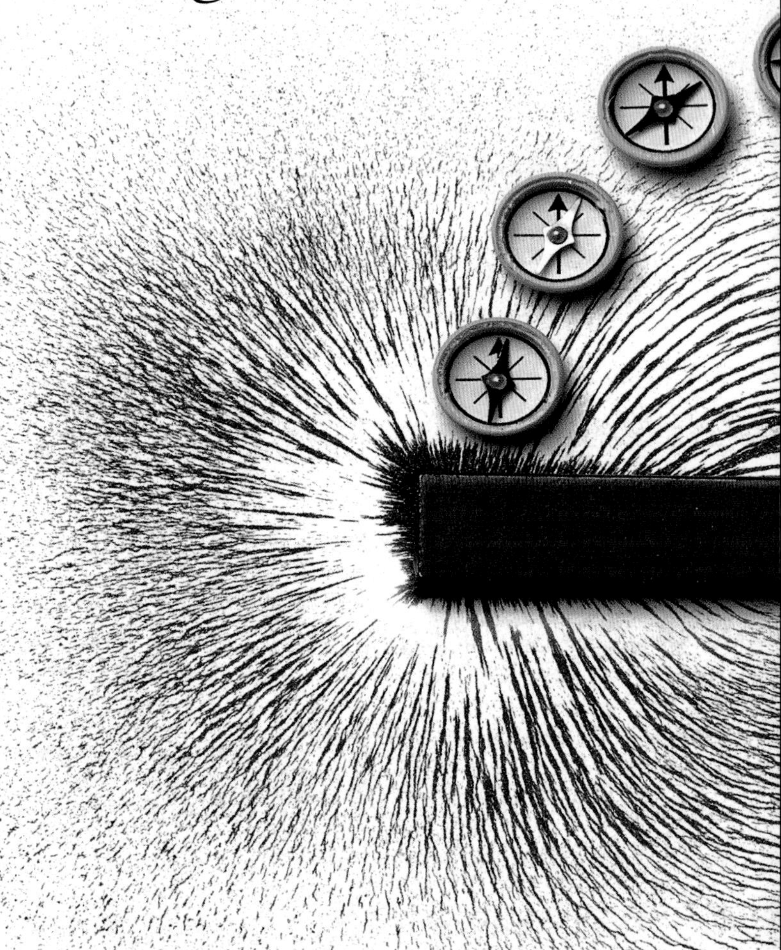

Magnets are surrounded by something called a *field of force*. It's invisible, but you can see it by scattering iron filings on a piece of paper and placing a magnet on top. The iron filings jiggle about until they align with the magnet's field. They also cluster around the poles, where the magnetic force is strongest.

FAQ

Why is Earth magnetic?

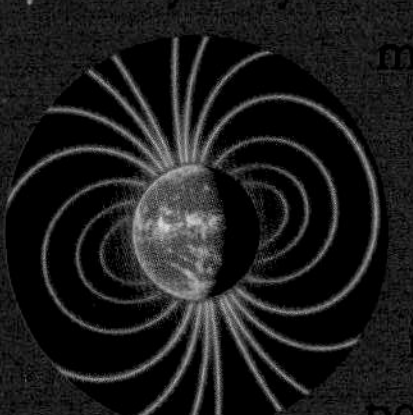

Earth is a giant magnet, but the reason is a mystery. Scientists once thought Earth must have a giant iron magnet as its core. Although the core is iron, it can't be a magnet because iron loses its magnetic powers above 760°C, and the core is at least 1,000°C. One possibility is that the molten core contains swirling electric currents that generate the magnetic field.

Why is Earth upside down?

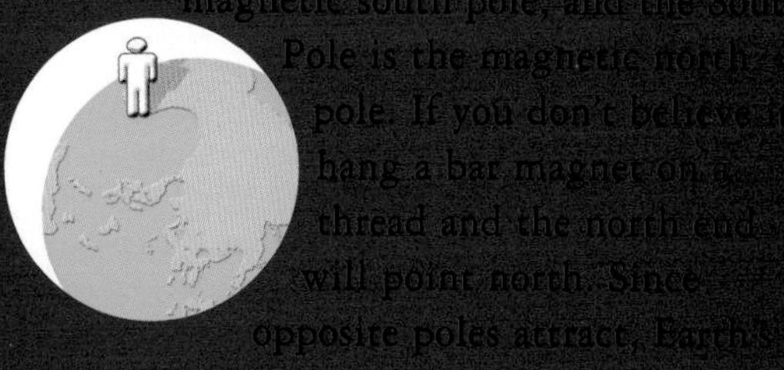

Earth's North Pole is actually the planet's magnetic south pole, and the South Pole is the magnetic north pole. If you don't believe it, hang a bar magnet on a thread and the north end will point north. Since opposite poles attract, Earth's North Pole must be a magnetic south pole.

What are the northern lights?

Earth's magnetic field shields us from a stream of electrons that pours out of the Sun – the "solar wind". Some of the electrons slip through the net, however. Drawn along magnetic field lines, they crash into the North and South poles and light up the night sky with magical colours.

Is the Sun a magnet?

The Sun is an even more powerful magnet than Earth. Its insides are always churning around, twisting its magnetic field into tangles. Vast storms of super-hot gas burst out of the Sun and flow around the twisted loops of its magnetic field, forming "solar prominences".

work?

Whenever an ELECTRON moves, it creates a magnetic field around it just like the field around a bar magnet. Every atom has electrons, so every atom is magnetic. Normally the atomic magnets in objects are jumbled up and their force fields cancel each other out. However, in some materials, like iron, the magnetic fields of atoms can line up together. Then the whole piece of iron acts as one magnet.

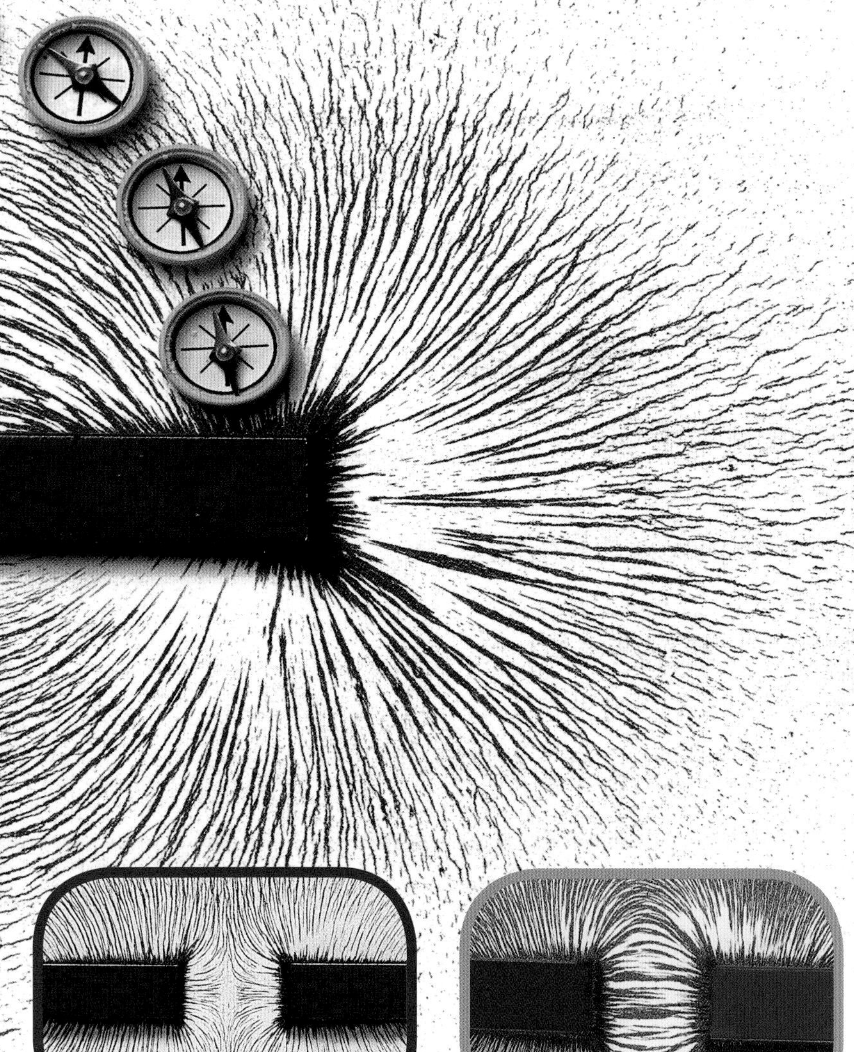

When the same poles of two magnets are pushed together, they repel each other. The field is weak between them and the lines of force curve away.

When opposite poles of two magnets are put together, they pull each other strongly. The force field is strong, and lines of force run between the magnets.

FAQ

Can electricity make magnetism?

Although magnetism and electricity seem like very different things, they are actually different aspects of the same force: ***electromagnetism***. The first inkling of a link came in 1802, when an Italian scientist discovered by accident that electric wires can swing compass needles round. The moving electrons in the wire were creating a magnetic field. The English scientist Michael Faraday went on to discover that the opposite can happen too: if you move a magnet next to a wire, the moving magnetic field generates electricity. Faraday had invented a way of creating electricity from movement. It was one the greatest scientific inventions ever, and today we get nearly all our electricity this way.

Make a compass

If you have a strong magnet, you can use it to make a compass. Stroke a steel needle with the magnet in one direction only for 15 seconds. Tape the needle to a piece of cork or styrofoam and float it in water. The needle will turn round to point north.

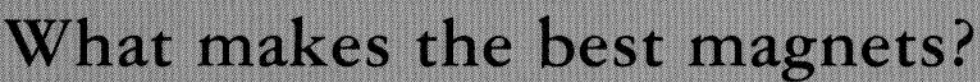

What makes the best magnets?

Neodymium magnets, which are made from a mixture of iron, boron, and neodymium, are up to 20 times more powerful than normal iron magnets. There are tiny neodymium magnets in earbuds – if you have a pair, try picking up paperclips or pins with them. A neodymium magnet the size of a penny can lift a 10 kg (22 lb) weight and will pick up metal objects through your hand.

Can you feel the *heat*?

Imagine you had a powerful microscope that could zoom in to look at atoms and molecules. You'd see the atoms and molecules jiggling about. We feel this jiggling motion as HEAT. The faster the atoms jiggle, the hotter they feel. ***Temperature*** is just a measure of how FAST the jiggling atoms are moving.

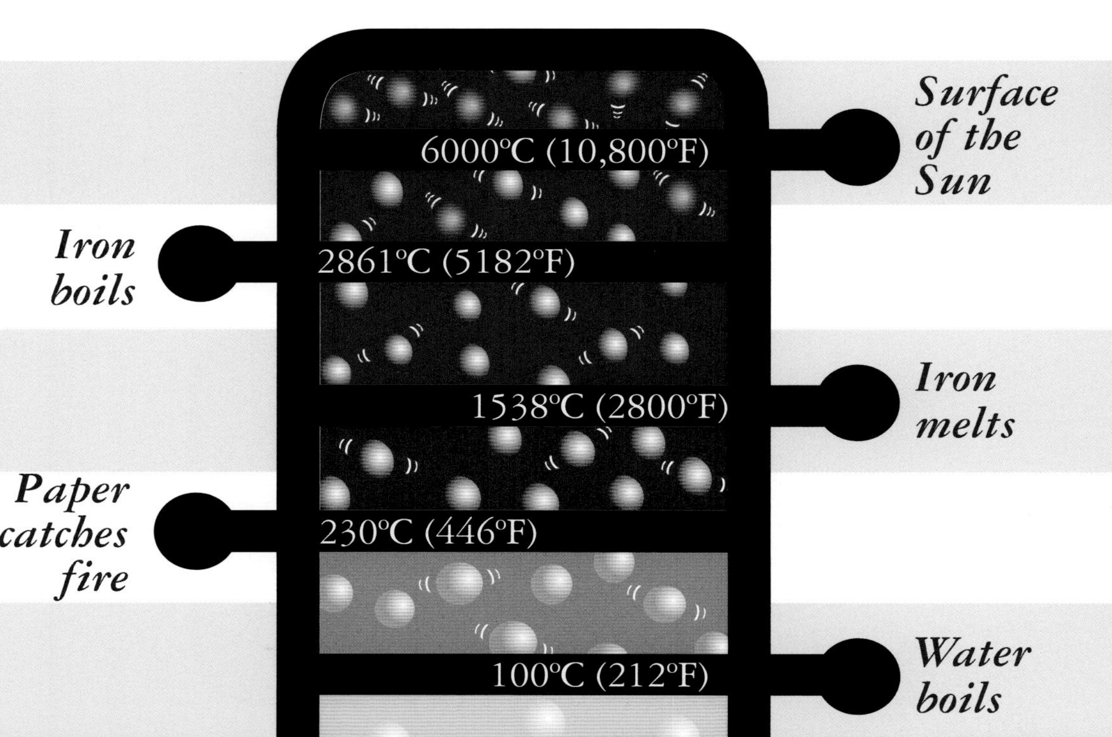

There are three ways that heat can move around:

CONDUCTION:

Molecules transfer their energy as they bounce into each other. When you hold a hot cup of coffee, heat enters your hand by conduction.

CONVECTION:

When air gets hot, it expands and becomes lighter. As a result it floats upwards, taking the heat energy with it.

RADIATION:

Hot objects give off heat as invisible rays, called infrared rays. You can feel the Sun's infrared rays as warmth on your skin.

FAQ

Why is metal cold and wood warm?

Metal often feels colder than wood, even when they are the same temperature. Why the difference? Metals are far better at ***conducting*** heat. They conduct heat away from your hand, cooling your skin.

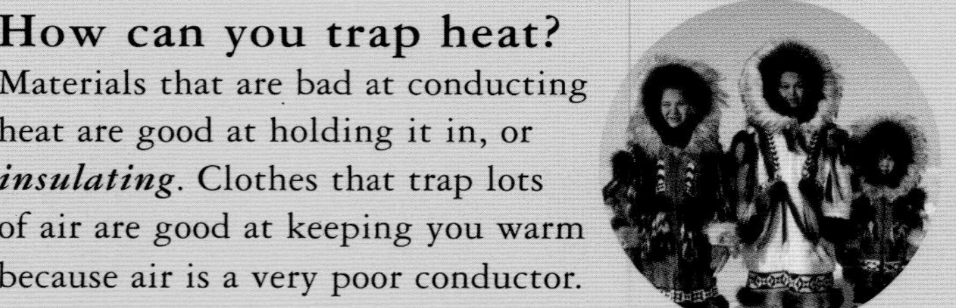

How can you trap heat?

Materials that are bad at conducting heat are good at holding it in, or ***insulating***. Clothes that trap lots of air are good at keeping you warm because air is a very poor conductor.

Hottest weather on Earth – 58°C (136°F)

Water freezes – 0°C (32°F)

Mercury freezes – –39°C (–38°F)

Coldest weather on Earth – –93°C (–136°F)

Air becomes liquid – –196°C (–321°F)

Air freezes – –219°C (–362°F)

Outer space – –270°C (–454°F)

Absolute zero – –273°C (–459°F)

Absolute zero...

is the lowest temperature possible, when atoms have no kinetic energy and stop moving. Actually, it's impossible to get all the way down to absolute zero, but you can get close. Weird things happen within a millionth of a degree of absolute zero. Atoms lose their individuality and slide into each other, forming a single atom-sized blob – a "Bose-Einstein condensate".

Can you see heat?

Although infrared rays are invisible to our eyes, special cameras can detect them and take "heat pictures", or thermograms. This thermogram of a plate of spaghetti shows hot areas as red and white but cold areas as blue.

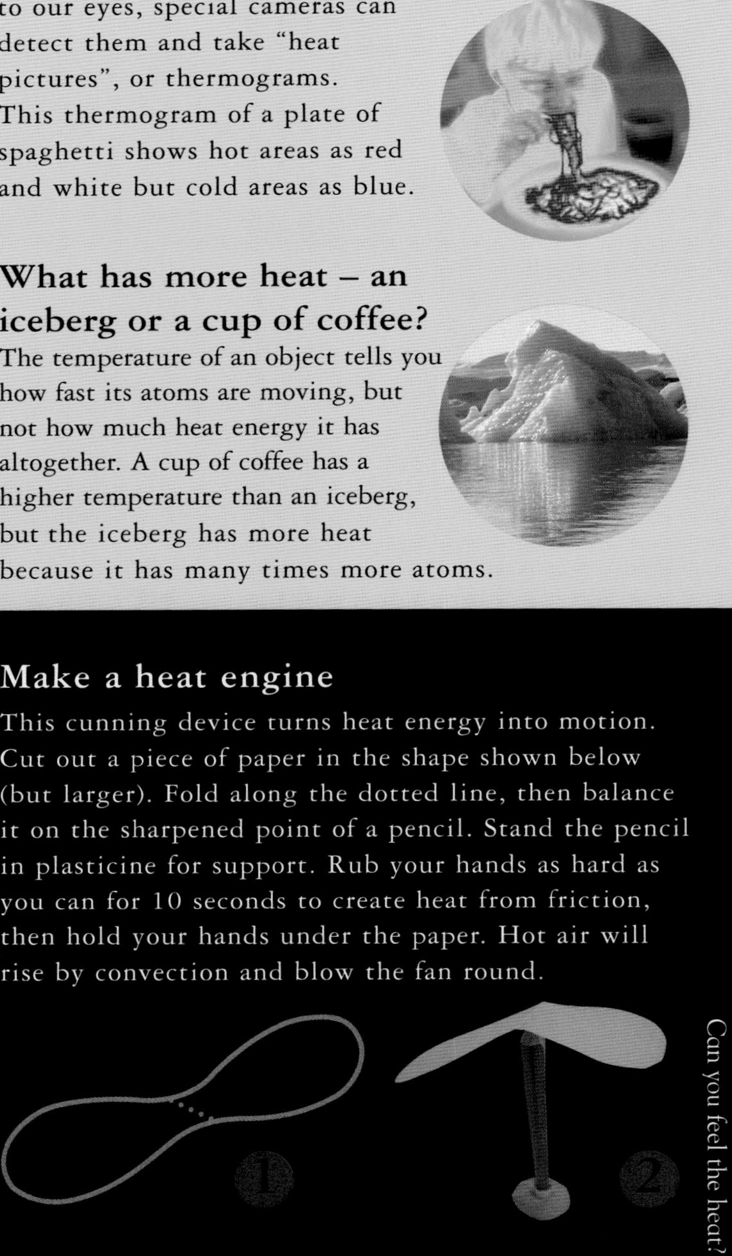

What has more heat – an iceberg or a cup of coffee?

The temperature of an object tells you how fast its atoms are moving, but not how much heat energy it has altogether. A cup of coffee has a higher temperature than an iceberg, but the iceberg has more heat because it has many times more atoms.

Make a heat engine

This cunning device turns heat energy into motion. Cut out a piece of paper in the shape shown below (but larger). Fold along the dotted line, then balance it on the sharpened point of a pencil. Stand the pencil in plasticine for support. Rub your hands as hard as you can for 10 seconds to create heat from friction, then hold your hands under the paper. Hot air will rise by convection and blow the fan round.

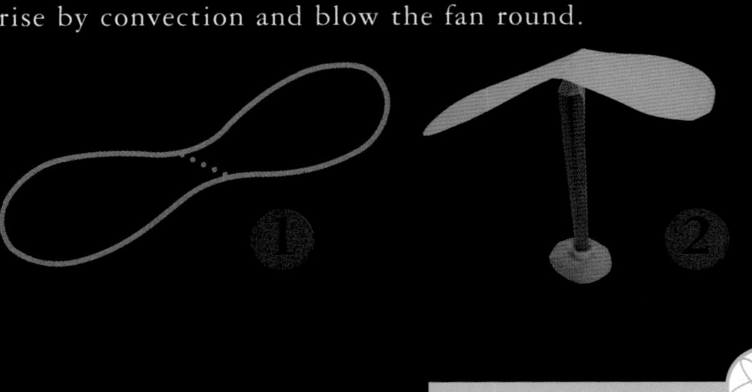

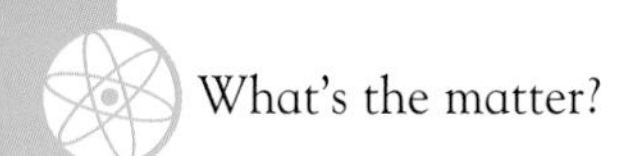

States of matter

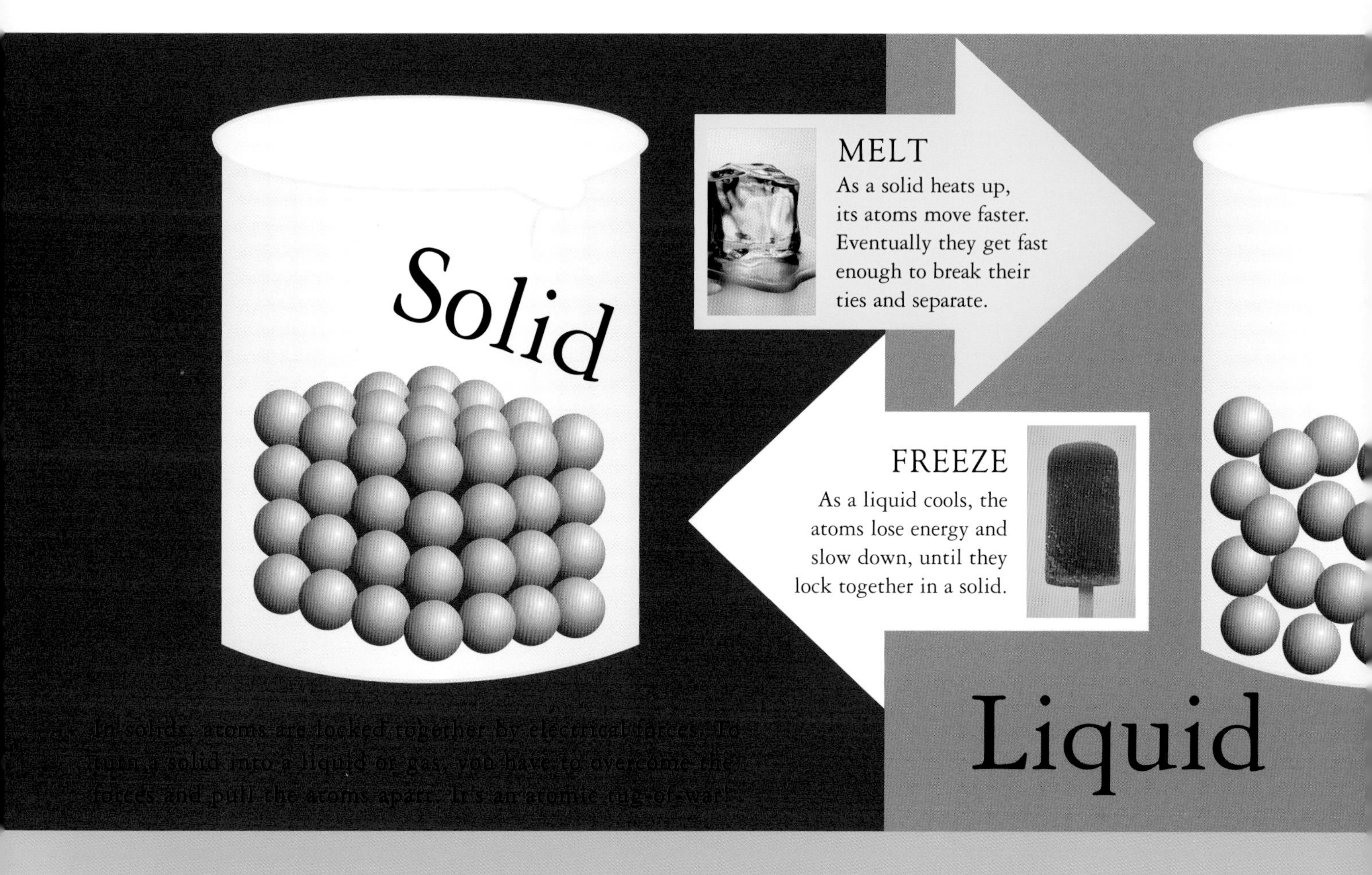

In solids, atoms are locked together by electrical forces. To turn a solid into a liquid or gas, you have to overcome the forces and pull the atoms apart. It's an atomic tug-of-war!

Cool or squeeze liquids to make solids

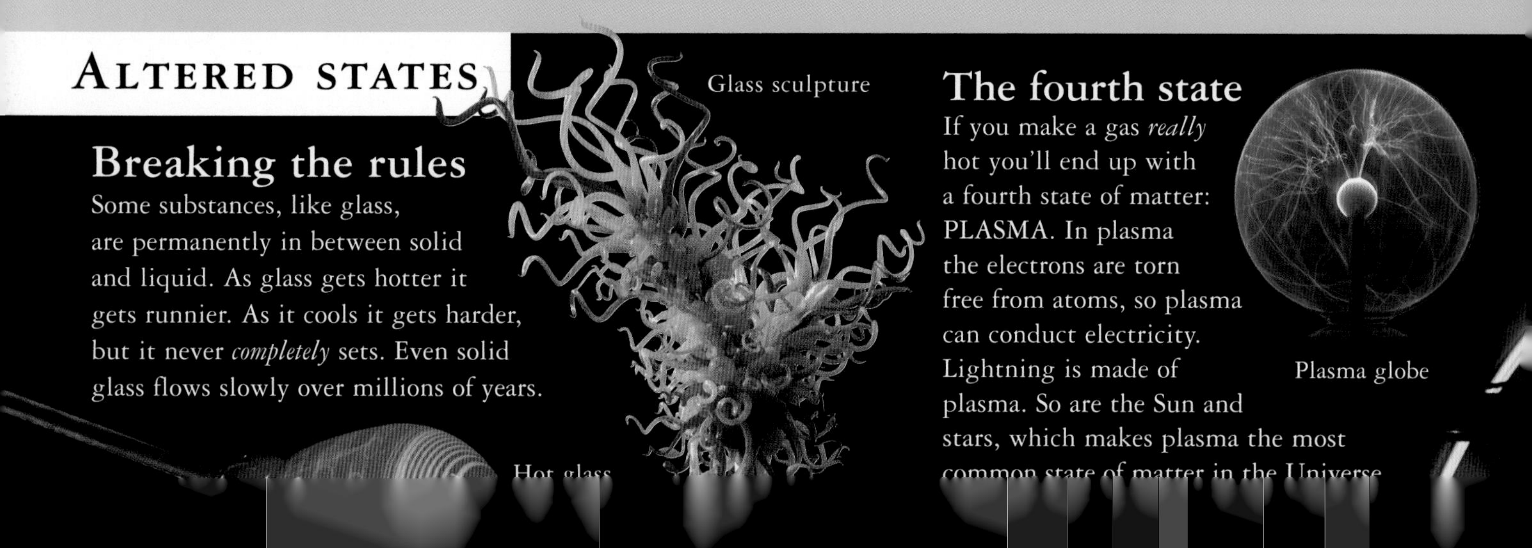

Altered states

Breaking the rules

Some substances, like glass, are permanently in between solid and liquid. As glass gets hotter it gets runnier. As it cools it gets harder, but it never *completely* sets. Even solid glass flows slowly over millions of years.

Glass sculpture

Hot glass

The fourth state

If you make a gas *really* hot you'll end up with a fourth state of matter: PLASMA. In plasma the electrons are torn free from atoms, so plasma can conduct electricity. Lightning is made of plasma. So are the Sun and stars, which makes plasma the most common state of matter in the Universe.

Plasma globe

Boil a glass of water until it ***disappears*** and you'll make enough steam to fill a room. Steam is water blown apart into a cloud, just as ice is water frozen tight into a block. Like ice, water, and steam, all matter can exist in three states: SOLID, LIQUID, and GAS. But these aren't the only states...

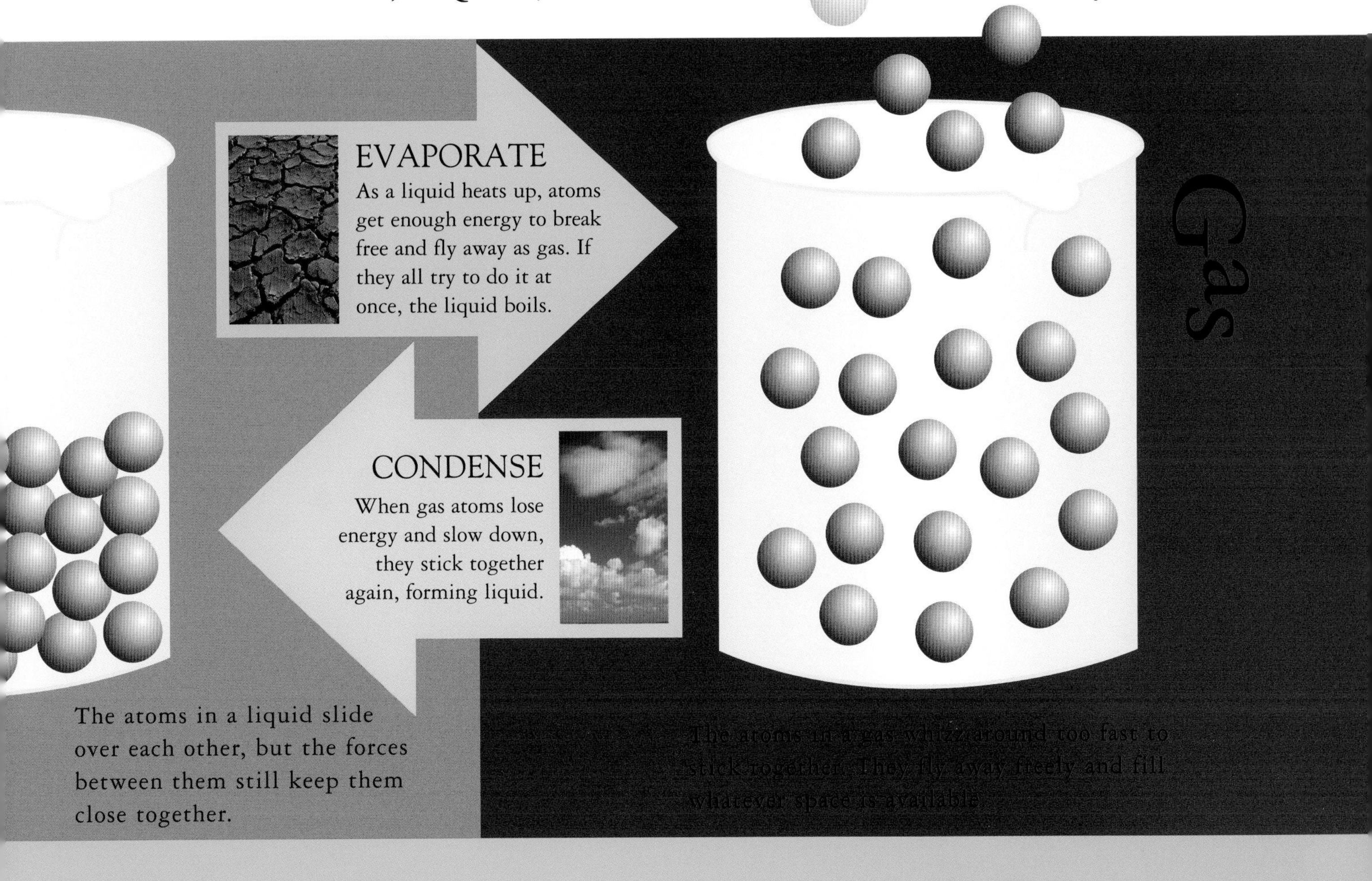

The atoms in a liquid slide over each other, but the forces between them still keep them close together.

The atoms in a gas whizz around too fast to stick together. They fly away freely and fill whatever space is available.

Heat or stretch liquids to make gases

Neon signs use plasma to produce light.

Star matter

Solids, liquids, gases, and plasma are not the only states of matter. In a neutron star, atoms are crushed so tight that they disintegrate into a sea of neutrons – a gigantic atomic nucleus several kilometres wide, with gravity a trillion times stronger than Earth's.

WHAT *shape is a*

Water is an amazing substance that covers 70 per cent of Earth's surface. What makes it so special are its ***molecules***. Each one has two tiny hydrogen atoms joined to one oxygen atom. This gives them a positive electrical charge at one end and a negative charge at the other. So they SNAP TOGETHER and stick to things like tiny magnets.

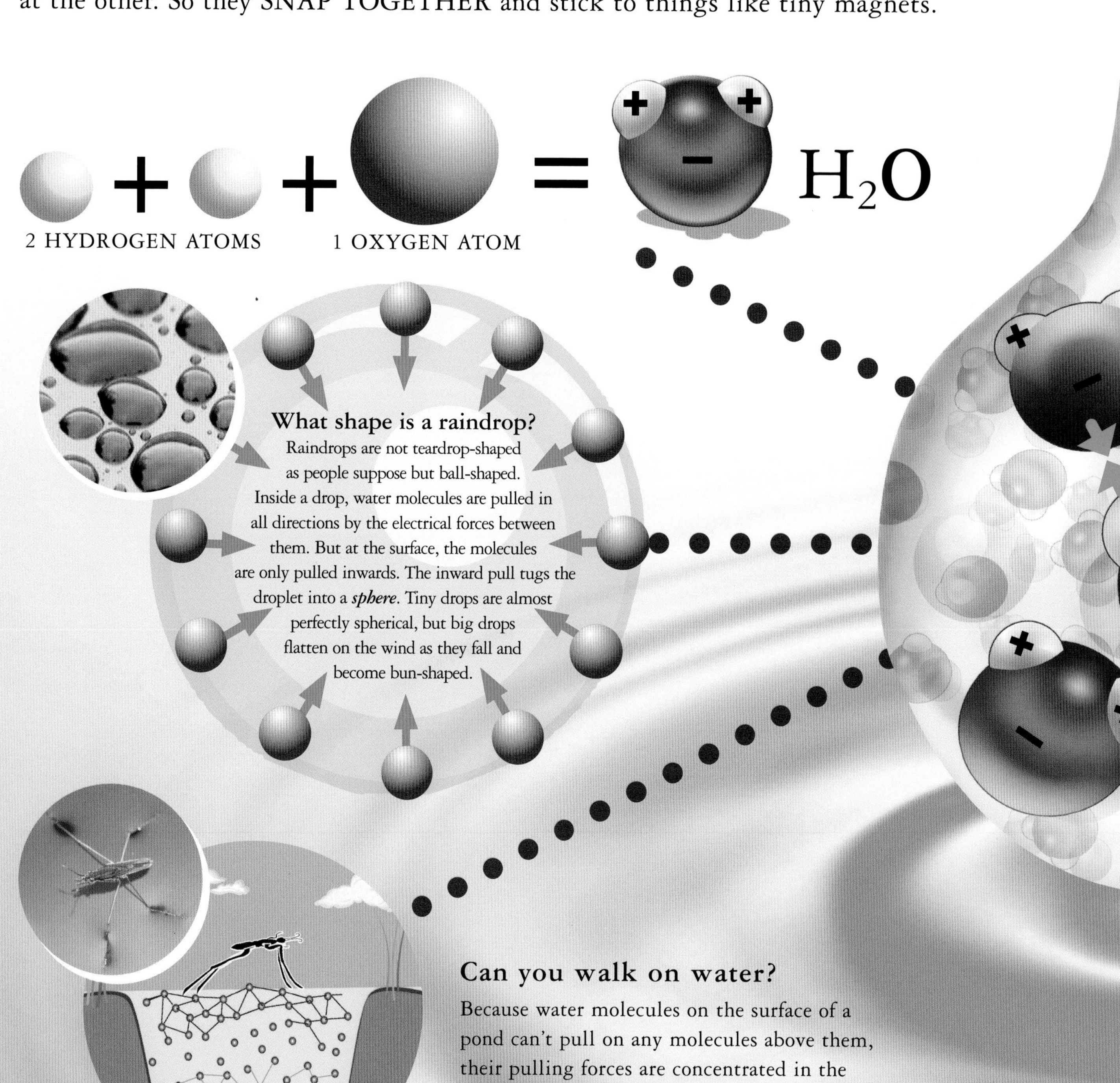

What shape is a raindrop?

Raindrops are not teardrop-shaped as people suppose but ball-shaped. Inside a drop, water molecules are pulled in all directions by the electrical forces between them. But at the surface, the molecules are only pulled inwards. The inward pull tugs the droplet into a ***sphere***. Tiny drops are almost perfectly spherical, but big drops flatten on the wind as they fall and become bun-shaped.

Can you walk on water?

Because water molecules on the surface of a pond can't pull on any molecules above them, their pulling forces are concentrated in the surface layer. The result is a tight, stretchy skin on the surface, which is strong enough for insects to walk on. This is called surface tension.

raindrop?

Why do icebergs float?

Most liquids get more compact as they freeze, but water is the opposite. Its molecules spread out as they solidify, making water expand by about 10 per cent as it freezes. As a result, ice is lighter than water and floats rather than sinking. With ice on the top, water underneath can stay warmer. This helps life survive in lakes and rivers in winter.

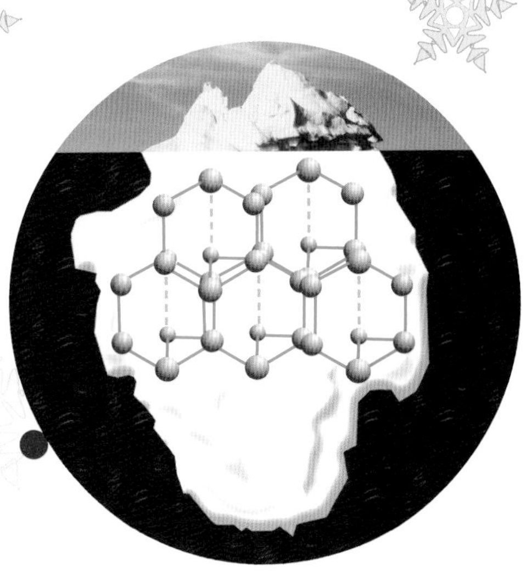

As water freezes into ice crystals, the molecules form an orderly pattern similar to the pattern of atoms in diamonds.

Float a paperclip

You can float a paperclip on water if you're careful. Place the paperclip on a fork and lower it very slowly into the water. The paperclip will be held by the stretchy skin of the water. Touch the water with a soapy finger to destroy the surface tension and make the paperclip fall.

AIR

WATER

Unpoppable balloon

Water can soak up and store heat better than almost any other substance. You can see its amazing heat-absorbing ability with this trick. Fill one balloon with air and another with water. Hold a candle beneath each one and see what happens. The air balloon heats up quickly and pops, but the water balloon survives much longer.

It takes lots of energy to break the bonds that hold water molecules together. This is why water takes a long time to get hot.

CAN YOU *walk on*

Some fluids are thin and fast-flowing, like water, but others are thick and slow, like honey. We call these thick fluids VISCOUS.

Stir a glass of water and it stays just as thin; stir a jar of syrup and it stays just as thick. This is how Isaac Newton thought all fluids should work, but there are some – such as custard and ketchup – that don't obey his rules. If you stir or shake them, the FORCE you apply changes their inner structure and makes them either runnier or more viscous. Fluids like this are called *non-Newtonian*.

Non-Newtonian?
How very dare you!

How does quicksand work?

Quicksand is a soupy mixture of sand and water that sometimes forms on beaches, riverbanks, over springs, or in marshes. It can look solid to the touch, but when you step on it you quickly sink in. Then, as you struggle to free yourself, the force of your movements makes the quicksand more viscous. Sand collects in a dense mass around your legs, ***trapping*** you.

How to sink

The worst thing to do in quicksand is to panic and struggle. Thrashing your arms and legs about not only makes the fluid thicken like cement but digs you in deeper. And the deeper you sink, the harder it is to get out.

How to swim

The secret to escaping is to stay perfectly still. The fluid then becomes less viscous and you automatically float to the surface, since quicksand is much more dense (and therefore easier to float in) than water.

Can you walk on custard?

Cornflour custard is a non-Newtonian fluid. If you apply a force to it, it turns solid. But does it harden enough to support the weight of a human being? In 2003, a team of researchers from the TV programme *Brainiac* decided to find out by filling a swimming pool with custard...

CUSTARD?

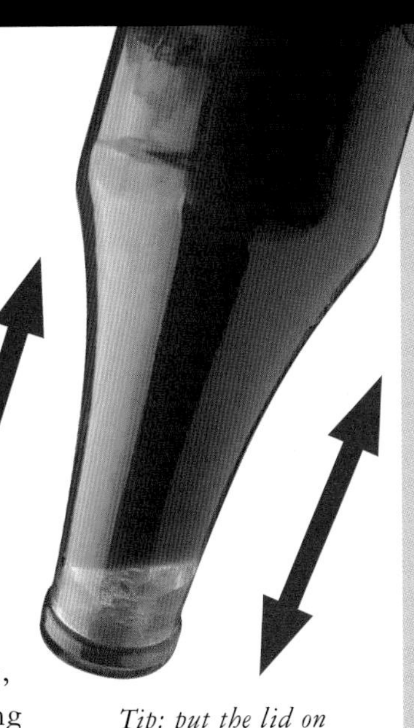

Why do you have to shake ketchup?

Ketchup is the opposite of quicksand – it gets runnier when you apply a force, which is why you have to shake the bottle to make it flow. Inside the bottle, ketchup is a jumble of solids and liquids held together by a kind of invisible scaffolding. When you shake it, the force smashes the scaffolding and the ketchup becomes less viscous.

Tip: put the lid on before you shake it.

Toothpaste

The technical term for liquids that become less viscous when you apply a force is ***thixotropic***. Toothpaste is also thixotropic. Take the cap off and it just sits there, peering at you from the nozzle. Start squeezing and the force breaks up its structure and makes it squelch out.

Make slime!

A good way to learn about non-Newtonian fluids is to make ***slime*** and explore its amazing properties. The recipe below tells you how to make cornflour slime, which is runny when you leave it alone but becomes so viscous when you apply a force that it TURNS SOLID!

Put 1½ cupfuls of cornflour in a mixing bowl. Slowly mix in a cupful or so of water, adding a little at a time and stirring. Add green food colouring to make the slime look like snot. Take a handful of slime and slap or squeeze it suddenly – it will turn solid. Let it relax for a few seconds and it will become liquid and will ooze beween your fingers.

Custard does support the weight of a human being, but only if you keep moving. The impact of each step hardens the custard below, which then forms a firm support. However, if you stop walking, the custard liquefies again and you slowly sink in.

HOW *does a balloon* BURST?

Ever wondered why balloons are so hard to blow up but so easy to burst? The physics of air and rubber has the answers.

Why do balloons stretch?

The secret to the stretchiness of balloons is in the molecules. Balloons are made of rubber, which has long, thin molecules like strands of cooked spaghetti. When you stretch rubber, the strands slide over each other and straighten out. When you let go, tiny forces between the molecules pull them back into a tangled clump.

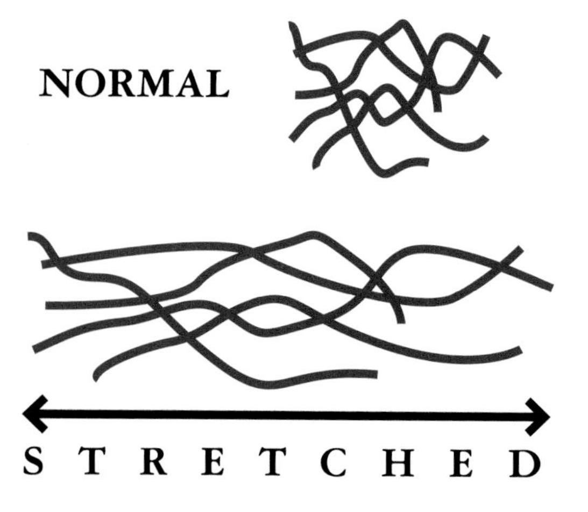

Why does it *BANG?*

A balloon takes a fraction of a second to burst. As it collapses, bits of rubber snap back to their original shape at more than twice the speed of sound and the air rushes out in a shockwave that you hear as a loud bang, like the crack of a whip.

Why is it hard to blow up?

It takes a lot of effort to blow up a balloon because you have to overcome the forces trying to pull the rubber molecules back together. The first breath is hardest because you have to straighten out billions of rubber molecules at once. After that, the molecules slide past each other a bit more easily as the rubber stretches, but they're still trying to pull themselves back. If you let go, the balloon will shrink suddenly and push out all the air. And that will make it fly out of your hand – a nifty demonstration of Newton's third law.

How does a balloon stay inflated?

Once you tie the knot, your breath is trapped in the balloon. A full balloon contains about 30,000 billion billion air molecules. Each one flies around at about 1600 kph (1000 mph) and bounces into other molecules or the wall of the balloon 5 billion times a second. The millions of billions of collisions with the rubber press against it, and this pressure keeps the balloon inflated.

What if you put a balloon in the fridge?

Try it. Stick a balloon in the fridge and see what happens. After about an hour, the air molecules will be flying about 50 kph (30 mph) slower and will hit the rubber with less force. But the force pulling the rubber molecules together will be just as strong, so the balloon will shrink until the forces balance again. If you warm the balloon on a radiator, the opposite happens and the balloon expands.

cold air

In a chilled balloon the air molecules move more slowly and hit the rubber with less force, so the balloon shrinks.

Now try this!

Can you pierce a balloon without bursting it?

Yes. If a balloon isn't overinflated, the rubber near the knot and around the top isn't fully stretched and will look thick and dark. Put a little grease or lip balm on the end of a skewer or a paper clip and gently push it into the dark rubber, twisting it until it pierces. With practice you can push a skewer right through both ends of the balloon.

What if you pierce it somewhere else?

Elsewhere the rubber is very taut. It's reached its "elastic limit", which means it will crack rather than stretch if the strain increases. Once tiny cracks appear, the pressurized air rushes out. Huge cracks race around the surface and the balloon bursts!

Balloon in a bottle

Here's a trick to baffle your friends. Show them an inflated balloon inside a bottle and challenge them to put their own balloon in another bottle and inflate it. When they try, the balloon will block the neck of the bottle and won't expand.

Here's the secret: put a drinking straw in the bottle at the same time to let air escape as you blow up the balloon.

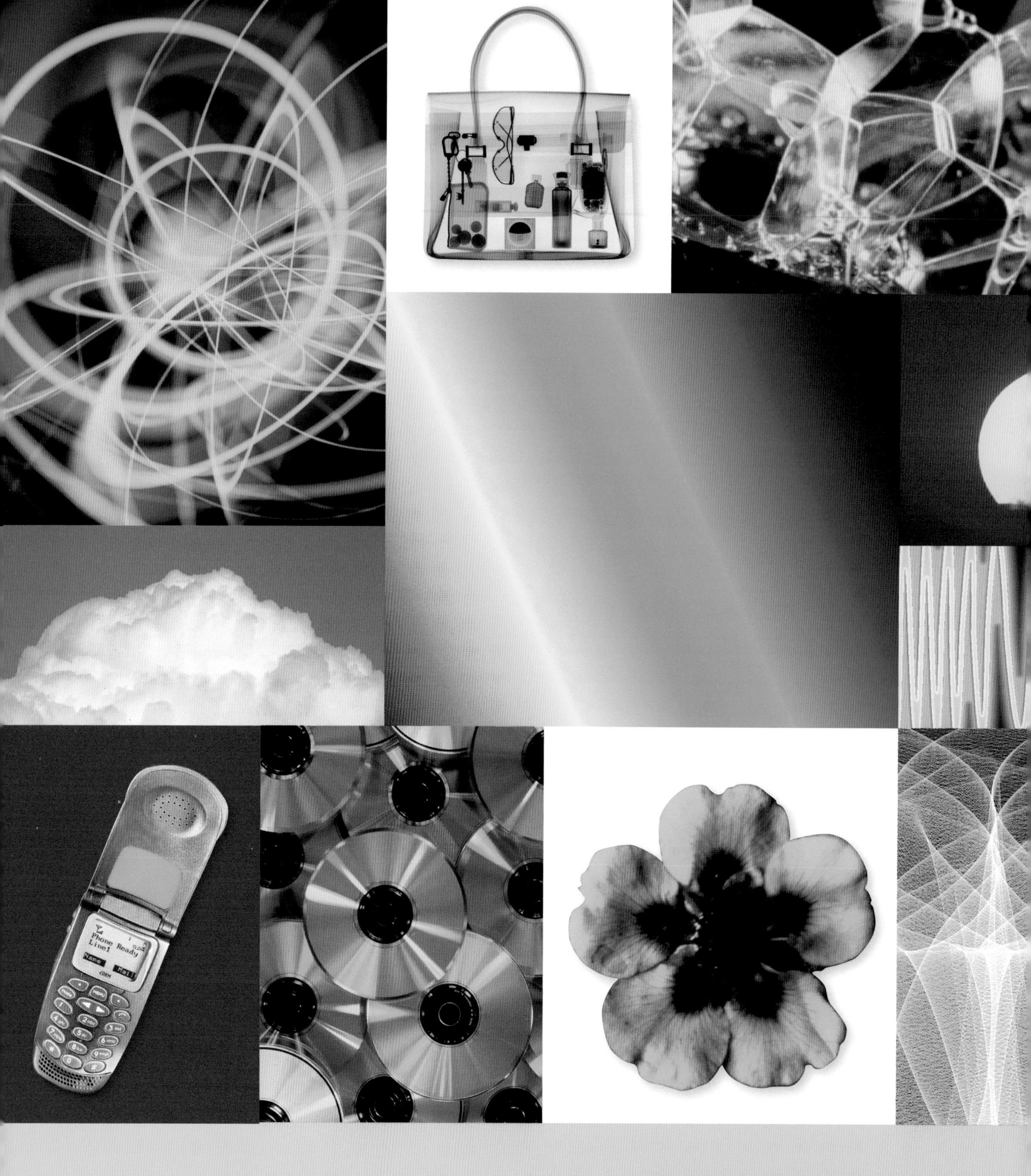

Can you *see* the LIGHT?

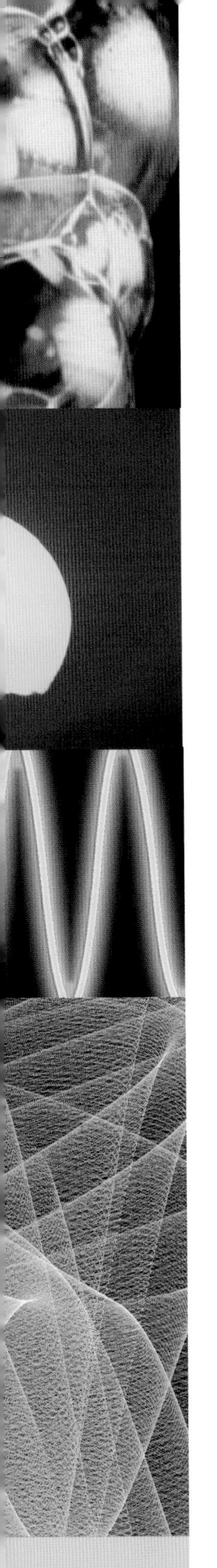

We all know that light is important – you wouldn't be able to read this page without it. You probably know that it's pretty quick too. Moving at a nifty

1 BILLION kph,

it's the fastest thing in the Universe.

But did you know that light could hold the secret to time travel? And it's impressively tricky stuff, sometimes acting as tiny particles and sometimes as ripples through the air, like radio waves.

IN FACT, LIGHT RAISES A LOT OF QUESTIONS. **How does it work?** Why do stars twinkle? ***Why are sunsets red?*** **And why did light cause the usually very clever Isaac Newton to stick a knitting needle in his eye?** Take advantage of the light around you right now to read on and find out.

PARTICLES

Is light made of *particles?*

Light is a riddle. It's the fastest thing in the Universe and weighs absolutely nothing. We see it all the time, but we can't see what it's made of. Sometimes it acts like billions of tiny particles flying past at amazing speed. At other times it seems like a wave that ripples through the air. So what is light?

Bouncy balls

Isaac Newton was one of the first scientists to think of light as particles. If it were a wave, he thought, it wouldn't always travel in straight lines and make sharp shadows – it would flow around objects the way ocean waves wash around rocks and sound waves bend through doorways. Newton also saw that light bounces off mirrors just as balls bounce off a wall, which made him think light must be made of tiny bouncy particles.

Lost in space

The particle theory seemed to explain how light could travel to Earth from the Sun. Particles can fly across empty space, but waves need something to ripple in. Sound waves are ripples in air, for instance, and ocean waves are ripples on water. If light is a wave, what is doing the rippling as it passes through outer space?

The mighty atom

After scientists discovered how atoms work, they found that an atom can spit out light particles one at a time. This proved that light really is made of particles, and they were named **photons**.

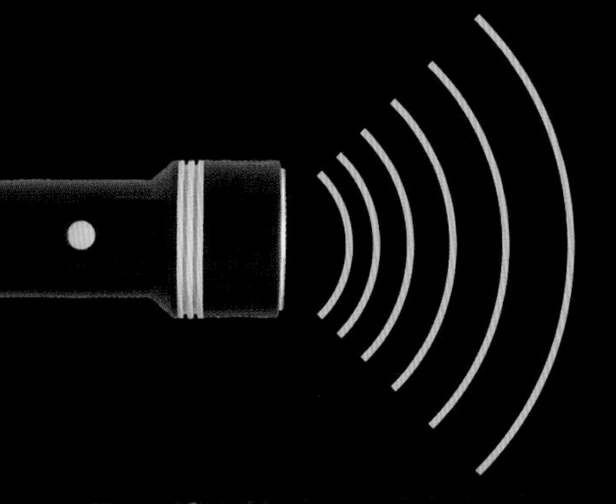

Is light made of *Waves?*

The answer to the riddle is that light is both things at once: a wave and a particle. It just depends on how you look at it. It isn't just light that has this peculiar ability to be two things at once. All the tiny particles in atoms and even atoms themselves can act like particles or waves depending on the circumstances.

Bendy light

The first inkling that light is a wave came from an Italian scientist called Grimaldi nearly 400 years ago. He made shadows with narrow beams of light and discovered that the shadows were slightly wider than they ought to be, as though the light had bent like a wave.

Ripples

Proof came about 200 years later. An English scientist called Thomas Young shone light through a grid of narrow slits. The light not only bent but formed ripply patterns similar to the patterns you see in a pond when two sets of ripples overlap. Young even worked out the length of the waves – they were less than a millionth of a metre.

Across the Universe

The wave theory explained why colours exist (each colour has a particular wavelength), but how could light waves travel in space? Scientists first thought space must be full of some invisible substance called "ether", but the true answer was much stranger. Empty space can contain invisible "force fields" exactly like the magnetic field around a magnet. Light is a ripple in these force fields.

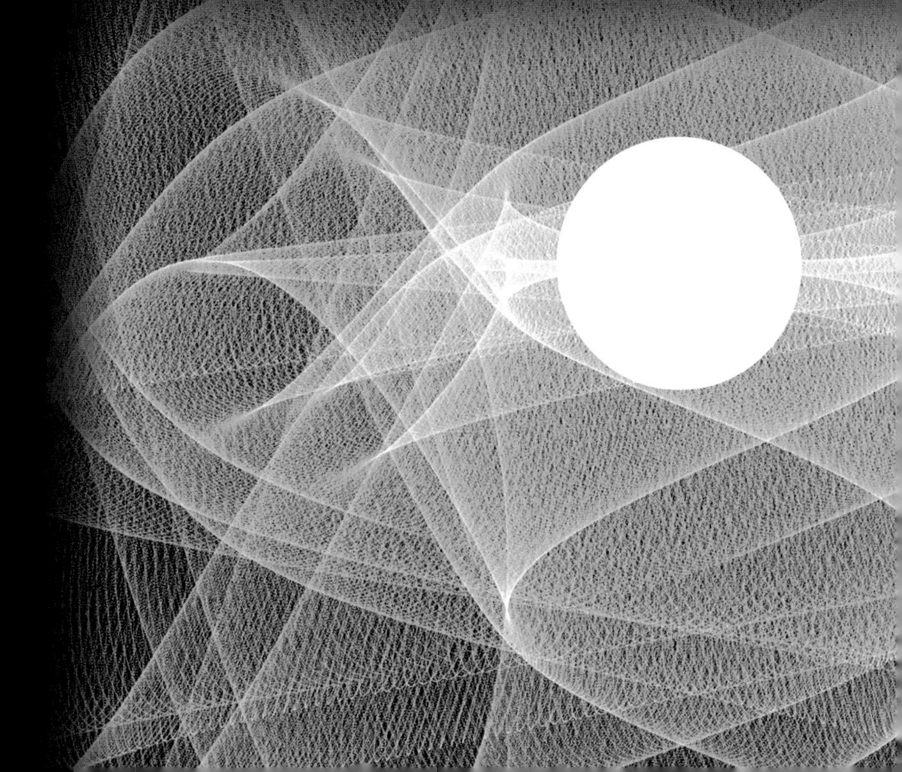

Particles or waves?

What *colour* is

Light that looks white

is really a mixture of every different colour, jumbled together so thoroughly that you can't see any of them separately. When sunlight shines through raindrops or bounces off the back of a CD, the hidden colours fly apart and reveal themselves in rainbow patterns.

How do we see colour? Our eyes contain three types of colour detector: one for red, one for green, and a third for blue. By combining them, we see every shade of colour. It's a clever system but it's easily fooled. A TV tricks you into seeing colours that aren't there. To make yellow, for instance, it mixes tiny green and red lights, triggering the same combination of cells in your eyes that yellow light would trigger.

Rainbows happen when the Sun is behind you and rain is falling in front. Sunlight enters each raindrop, bounces inside, and flies back out the front, spreading into colours like light from a prism. Rainbows are arched because you only see colours where light hits at a certain angle. If the ground wasn't in the way, the rainbow would be a circle.

Why do colours exist? Colours exist because light waves can be different lengths.

If the waves are LONG, we see them as red.

If the waves are SHORT, we see them as blue.

Light waves are very short, even in red light. About 2000 waves stretching end-to-end would fit along a millimetre.

When colours are arranged in order of wavelength, they make a pattern called a *spectrum*. The spectrum has seven main bands that blend into each other gradually, creating countless different colours. The human eye can see as many as **10 million** colours, including some that aren't in the spectrum at all, such as brown and magenta.

Learn the phrase to *remember*

RED ORANGE YELLOW

Richard *of* *York*

LIGHT?

Who discovered the colours in light?

On a sunny day in 1665, when Isaac Newton was only 22, he shut himself in a darkened room in his mother's farmhouse in England. He let a sliver of light through a chink in the curtains and put a triangular wedge of glass – a "prism" – in the beam of light. The beam spread out into a spectrum of colours. Scientists already knew about this beautiful effect, but they thought the colours came from the glass. Newton proved otherwise. He put a second prism in the spectrum and focused the rays back together. Lo and behold, a spot of white light appeared on the wall.

Within weeks of his discovery, Newton was *famous.*

Sight for sore Isaac

Spurred on by his success with prisms, Newton carried out more experiments. One was very stupid. Newton thought the human eye might work like a prism, splitting light into the colours we see. To test this theory, he shoved a knitting needle into the back of his eye and squidged his eyeball to see if colours appeared. They didn't. The theory was wrong, and Newton ended up with a nasty eye infection that nearly blinded him.

FAQ

Why do diamonds sparkle?

Diamonds can split white light into colours even more effectively than a glass prism can. A well-cut diamond also makes the light bounce around inside. That's why diamonds sparkle with brilliant, fiery colours.

Make a spectrum

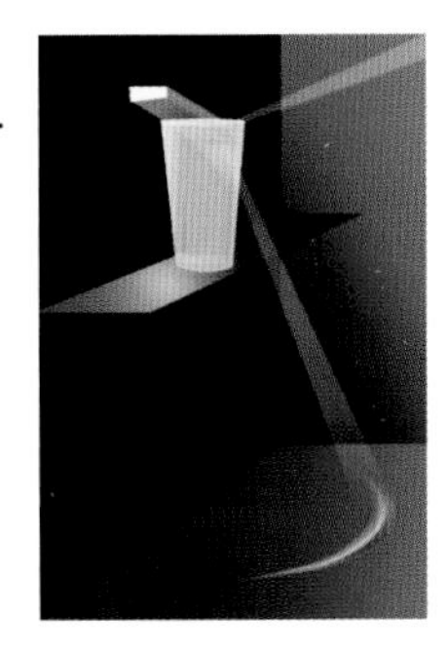

Make a spectrum with a glass of water. Stand a glass of water on a sunny windowsill. Cut a slit in a large piece of card and place the card behind the glass so the sunlight shines through the water. As the white light crosses the water surface, it will bend and split to form a spectrum.

Add colours together

You can add colours together by making a colour wheel. Use paint or a computer to create a coloured disc like the one below. Glue the disc to card, push a sharp pencil through the middle, and spin it. If you get the colour combination just right, the disc will turn white!

the order of colours of the spectrum

GREEN **B**LUE **I**NDIGO **V**IOLET

gave battle in vain

Can you see rainbows in *bubbles?*

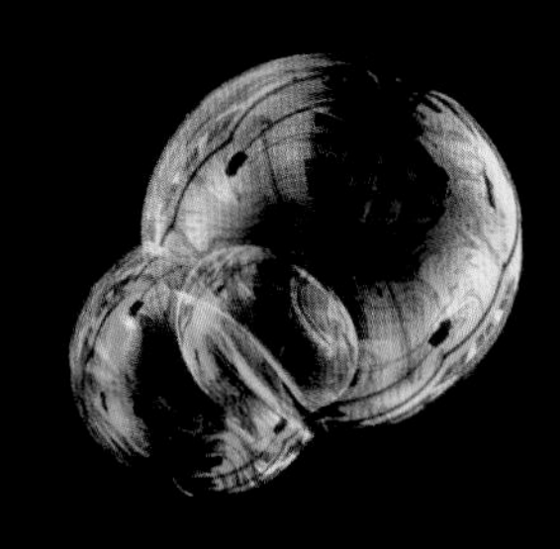

A bubble is

like a balloon of air, but its skin is a film of liquid rather than rubber. The film has three layers: an inner and an outer layer of soap molecules, and a layer of water molecules sandwiched between. The water molecules pull on each other, creating a force called surface tension, which holds the bubble together. The film is less than a thousandth of a millimetre thick, which is about the same as the wavelength of light.

soap
water
soap

When light

waves reflect off a bubble, they crash together and interfere. You can see a similar effect by throwing pebbles in a pond. One pebble makes a set of circular waves that spread out neatly, but two pebbles make two sets of waves that interfere. Where the peaks of two waves combine, they make a bigger wave. If a peak of one wave joins with the valley of another, the two waves cancel out. Light waves do just the same thing.

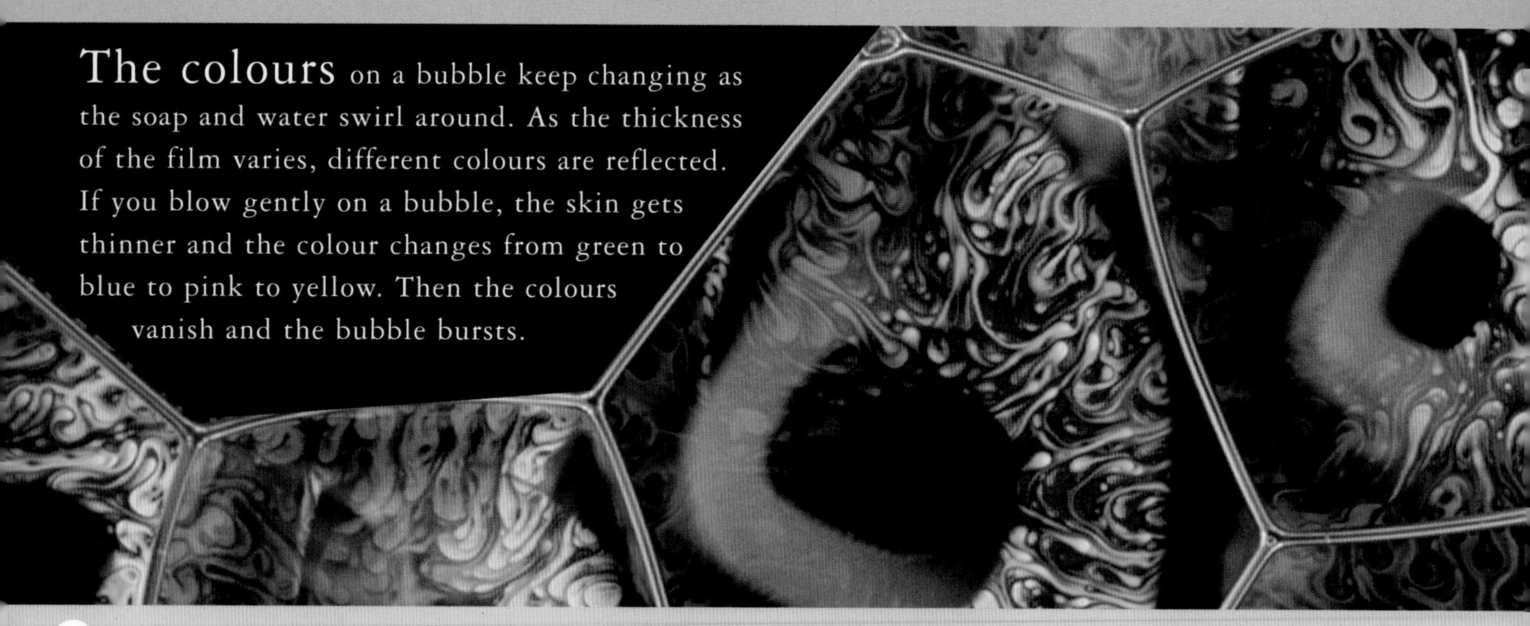

The colours on a bubble keep changing as the soap and water swirl around. As the thickness of the film varies, different colours are reflected. If you blow gently on a bubble, the skin gets thinner and the colour changes from green to blue to pink to yellow. Then the colours vanish and the bubble bursts.

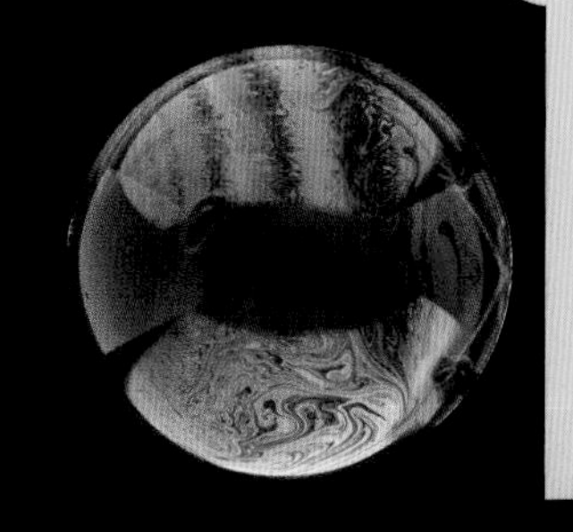

Peer at a soap bubble and you'll see rainbow colours swirling all over it. If you blow on the bubble, the colours will spin and change...

... but where do the *colours* come from?

Light reflects

off the inner and outer surfaces of a bubble film, making two sets of waves that interfere. Where the bubble's skin is just thick enough, certain wavelengths bounce back in step. Those are the colours you see. Other wavelengths bounce back out of step and cancel out, so those colours disappear.

white light

green waves in step

white light

red waves cancel out

Now TRY THIS!

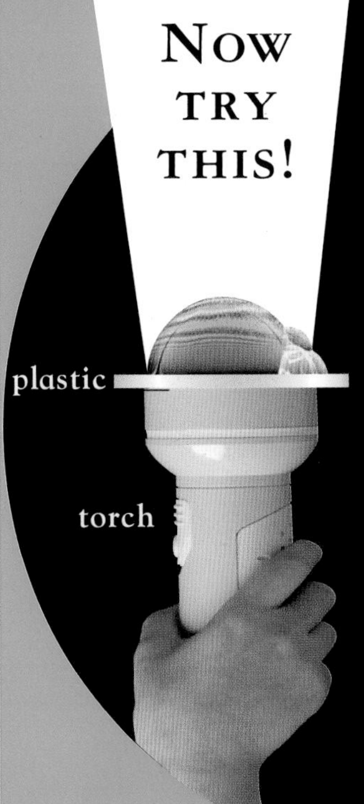

Make a **bubbularium** to light up bubbles from inside! Tape a square of clear, stiff plastic onto a torch. Put a spoonful of bubble liquid on the plastic and use a straw to blow bubbles in it. Turn out the lights, switch on the torch, and hold the bubbularium above eye level.

Tip: add gelatin or sugar to the bubble liquid to make stronger bubbles with brighter colours.

When is light invisible?

Suppose you used a pair of tweezers to stretch waves of light. As the waves got longer, they'd change colour. Then they'd become **invisible**, because our eyes can only see certain wavelengths. There are many types of light with wavelengths too long or short for us to see. Along with visible light, they make up the vast "electromagnetic spectrum".

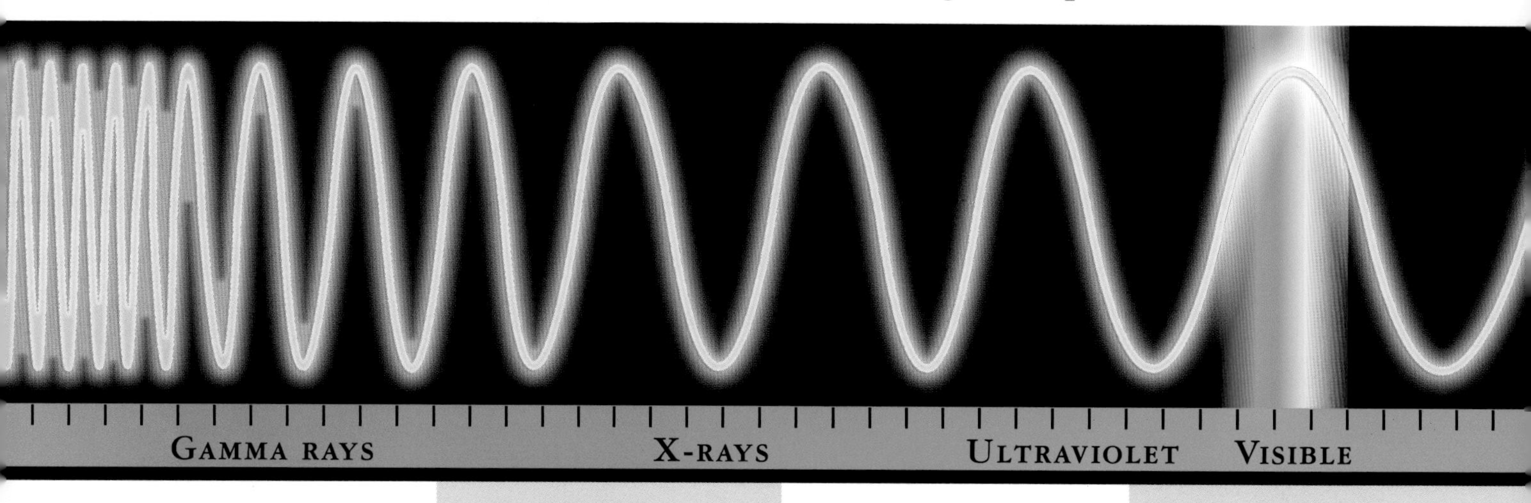

Gamma rays are the most dangerous electromagnetic rays. Their waves are even smaller than atoms but are packed with energy and can zip right through solid objects and kill living cells. They are given off by nuclear bombs as "radioactivity" and are used to destroy cancer.

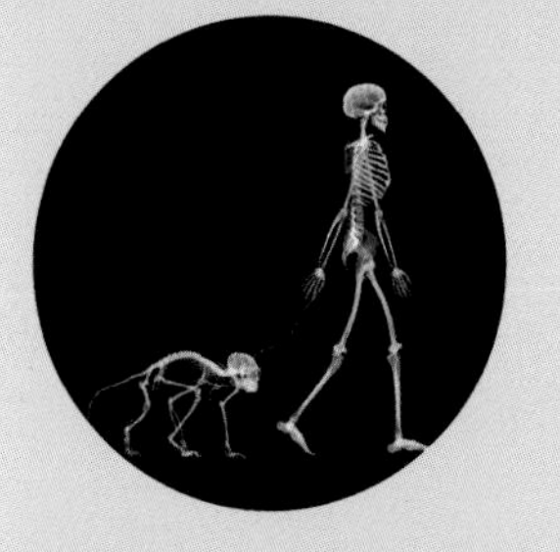

X-ray waves are about the size of atoms and contain lots of energy, but are less harmful than gamma rays. They pass through most parts of the human body but are blocked by bones, teeth, and metal, which makes them ideal for seeing inside things.

Ultraviolet (UV) rays come from the Sun but we can't see them, though bees, birds, and butterflies can. Some wavelengths of UV light can pass deep into human skin and damage the living cells inside, causing sunburn, cancer, or wrinkles that make people look old.

Visible light is the only type of electromagnetic radiation that our eyes can detect. It comes in a range of wavelengths, which we see as colours. These wavelengths are ideal for vision because sunlight is full of them and they bounce off objects rather than passing through.

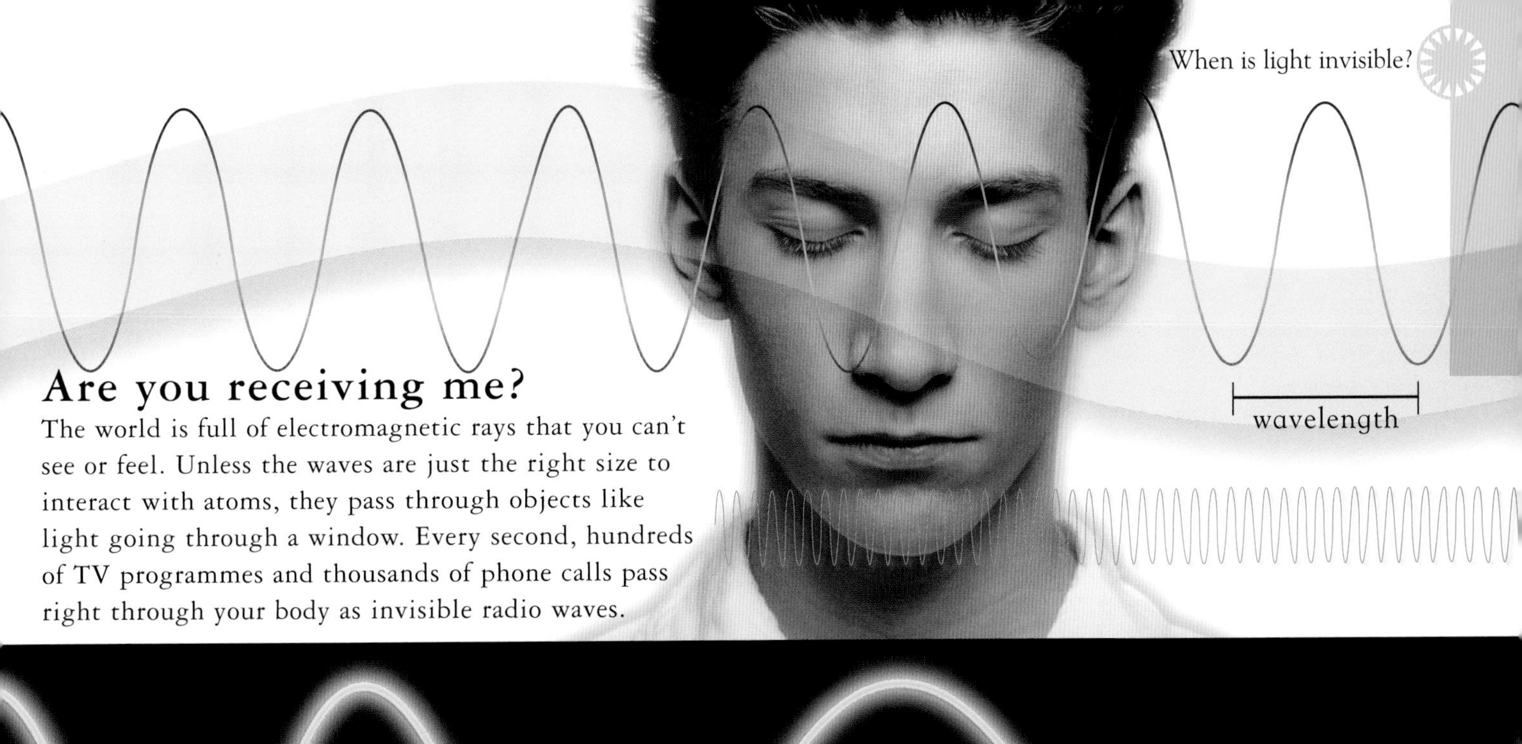

Are you receiving me?

The world is full of electromagnetic rays that you can't see or feel. Unless the waves are just the right size to interact with atoms, they pass through objects like light going through a window. Every second, hundreds of TV programmes and thousands of phone calls pass right through your body as invisible radio waves.

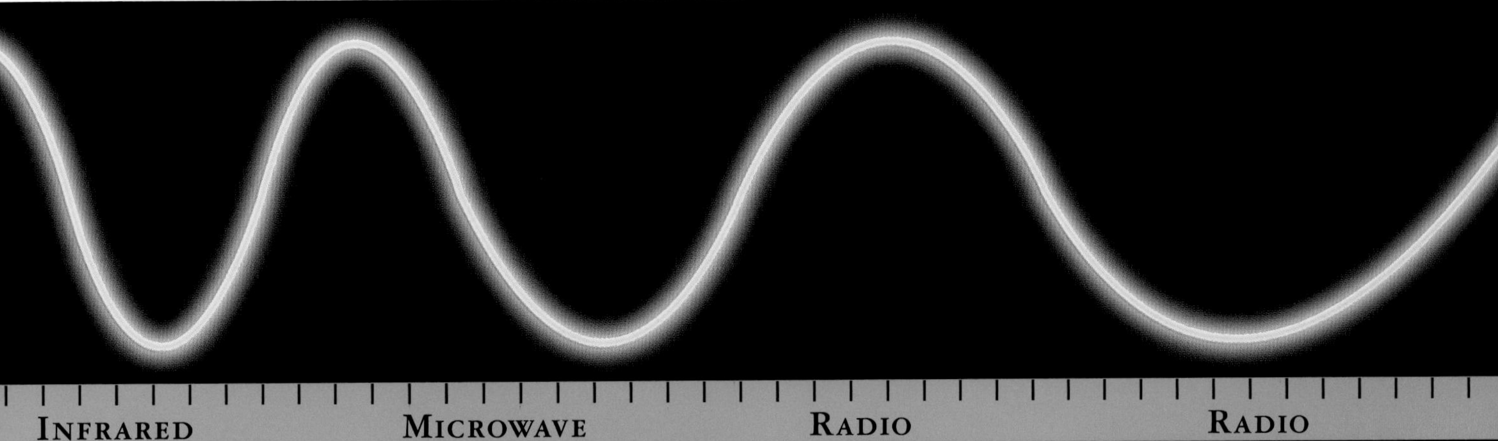

INFRARED

MICROWAVE

RADIO

RADIO

Infrared rays carry heat. Although they're invisible, you can see them if you wear night-vision goggles, and you can feel them on your skin when you warm your hands by a fire. Infrared waves vary in size from microscopically tiny to the size of a pinhead.

Microwaves are very short radio waves. They vary from the size of a pinhead to the length of your arm. Microwave ovens flood food with waves about 12 cm long. These wavelengths heat up water molecules but pass through things like glass and plastic.

Mobile phones and Wi-Fi gadgets send their signals as short radio waves, which vary from the size of microwaves to a few metres long. Short radio waves travel best in straight lines and are not good at bending round obstacles.

TV and radio stations broadcast in longer radio waves, from a few metres long (TV and FM radio) to hundreds of metres long (AM radio). Very long waves bend around obstacles and curve through the atmosphere, which allows them to travel around the world.

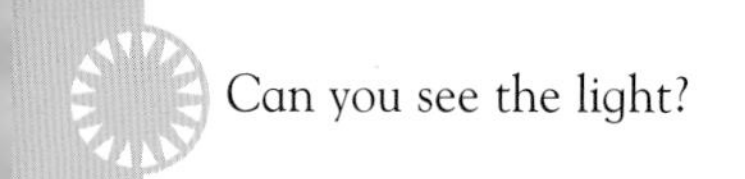

Why is the sky BLUE?

The *colour* of the sky comes from the *air*, which looks blue when sunlight hits it. If you were on the Moon, which has no air, the sky would be jet black. Earth, however, is covered by a thin blanket of air – the atmosphere.

When sunlight reaches Earth after its long journey from the Sun, it passes through the atmosphere. Remember that white sunlight is actually a mixture of all the colours of the rainbow. Some colours, such as red, cut straight through air without a problem. But other colours – particularly blue – crash into air molecules and bounce off in new directions. When you look up, you see this scattered light as a blue glow coming from the air.

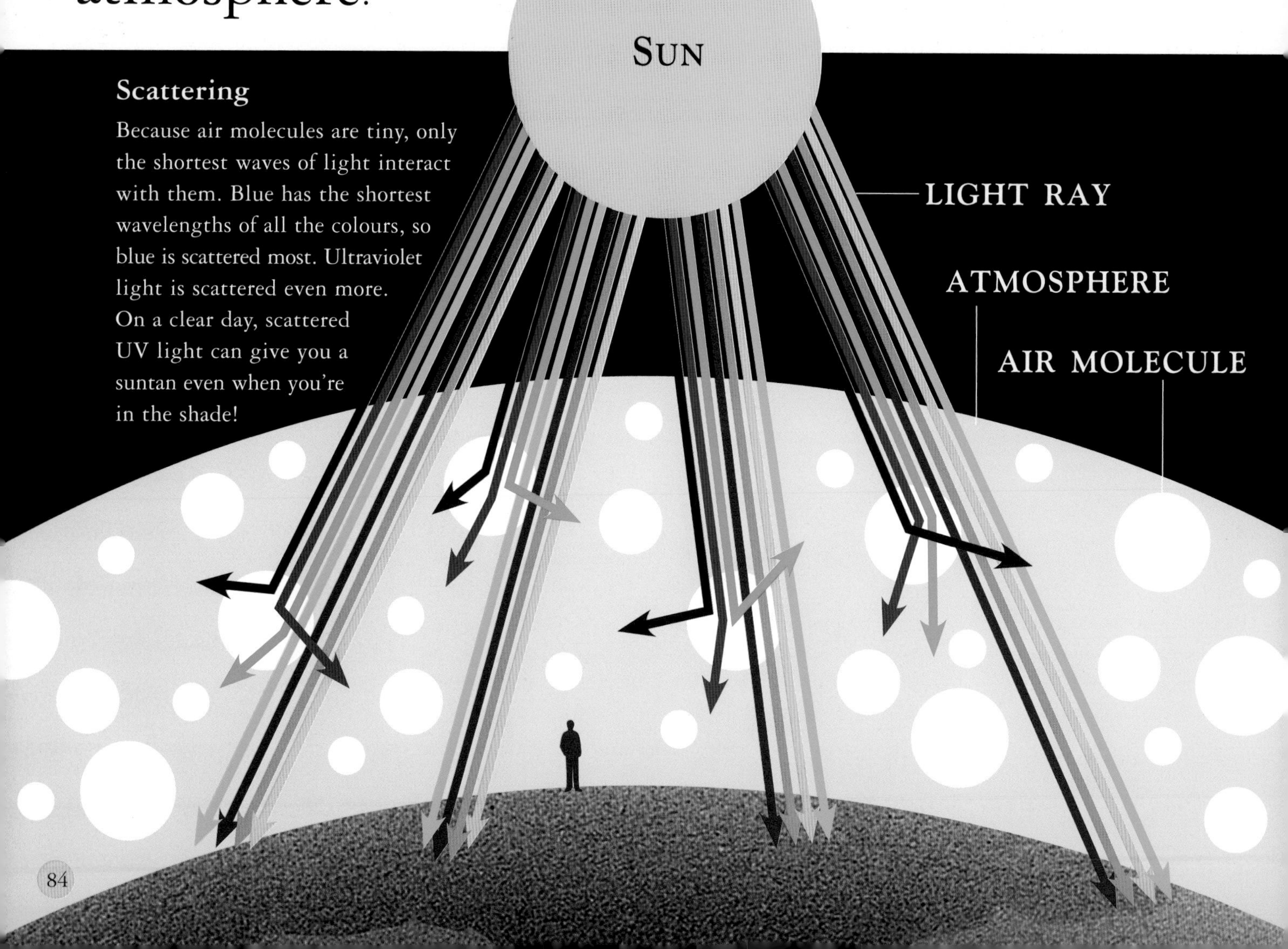

Scattering

Because air molecules are tiny, only the shortest waves of light interact with them. Blue has the shortest wavelengths of all the colours, so blue is scattered most. Ultraviolet light is scattered even more. On a clear day, scattered UV light can give you a suntan even when you're in the shade!

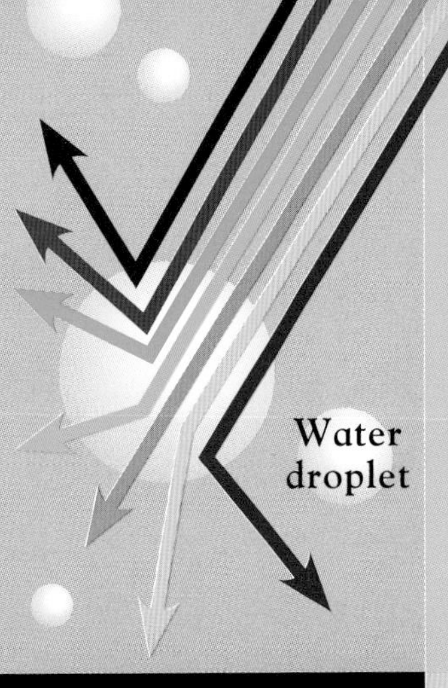

Why are clouds white?

It's not just air molecules that scatter light. The water droplets in clouds scatter light too, but they are much bigger than air molecules and so can scatter long as well as short wavelengths. Since all the different wavelengths add up to make white, the light we see from clouds is white.

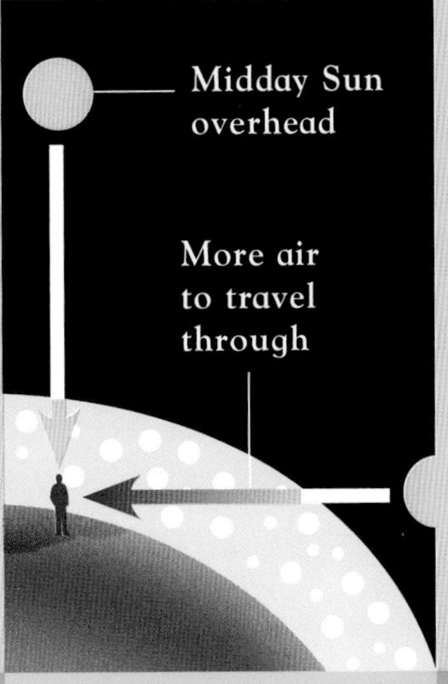
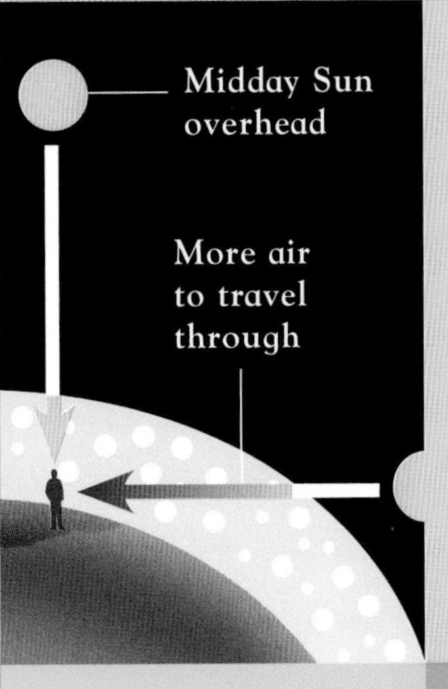

Why are sunsets red?

As the Sun sinks at the end of the day, its light has to travel through a thicker and thicker slice of the atmosphere. The air scatters more and more of the short wavelengths away, leaving just long wavelengths like red and orange. And that's why sunsets are red. Very fine dust, such as salt particles over the sea or ash from volcanoes, can make the sunset even redder.

Why is the sea blue?

A glass of water is colourless, but water in the sea is deep blue on a sunny day. The colour of water is caused by a process different from scattering. Water absorbs long wavelengths of light, such as red, but lets short wavelengths like blue pass through and bounce around. So light that passes through water is blue. In very deep water, the blue is absorbed too and the sea turns black.

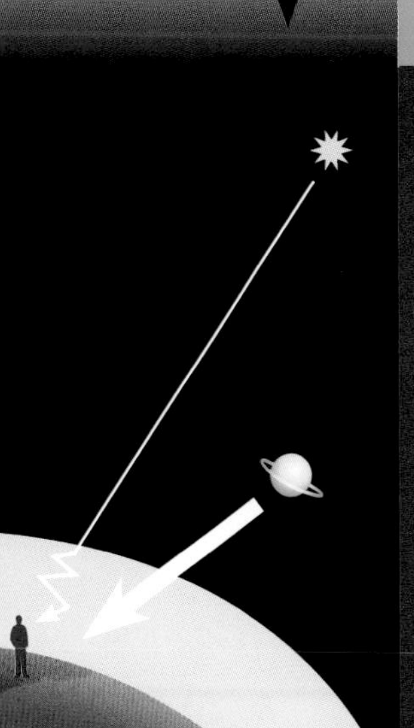

Why do stars twinkle?

Look at the stars and you'll notice they shimmer and change – they twinkle. But if you're lucky enough to see a planet, it won't twinkle at all. Why the difference? Stars are millions of times further away, so we see them as tiny dots of light. The beam of light from a star is so thin that pockets of hot and cold air in the atmosphere can bend it, making the star twinkle.

How *fast* is Light?

Light is the **fastest** thing in the Universe. It travels at about 1 billion kph (670 million mph), which is 10 million times faster than the motorway speed limit and 40,000 times faster than the Space Shuttle. At that speed, a beam of light can zip around the world *seven* times in just one second.

Why is light weird?

According to the laws of physics, there's something odd about light. Imagine you're chasing a car that's moving at 30 kph. If you ran behind it at 29 kph, the car would only be going 1 kph faster than you. You could almost keep up. The weird thing about light is that this doesn't happen. No matter how fast you chase light, it always moves away from you at exactly the same speed. Even if Superman was flying at 999,999,999 kph behind a beam of light, the light would still speed away from him at a full 1 billion kph. He might as well be *standing still*.

No matter how fast you fly, light *always*

Surely that's impossible?

No. The first person to realize that light really does travel in what seems an impossible way was Albert Einstein. Einstein worked out that if light doesn't change its relative speed, then time and space must be shrinking and stretching to make up for it. That means that as Superman gets close to the speed of light, his body shrinks and time slows down.

Can light slow down?

Light only reaches its top speed in the emptiness of space. If something gets in the way – like air, water, or glass – it slows down. The sudden change of speed also bends light, and this is why straight objects placed in water often look crooked. The bending effect is called **refraction**. If it didn't happen, we couldn't make telescopes, cameras, magnifying glasses, or spectacles.

Albert Einstein
(1879–1955)

rushes away at a billion kilometres an hour

Can you *travel*

1% *of light speed*

Imagine you're in a racing car that can go as fast as you like. Your challenge is to reach the speed of light and see what happens. You step on the accelerator and push the speed up to 10 million kph – just 1% of light speed. So far, things seem pretty normal. So you put your foot down and speed up some more...

If you drove at the speed of light

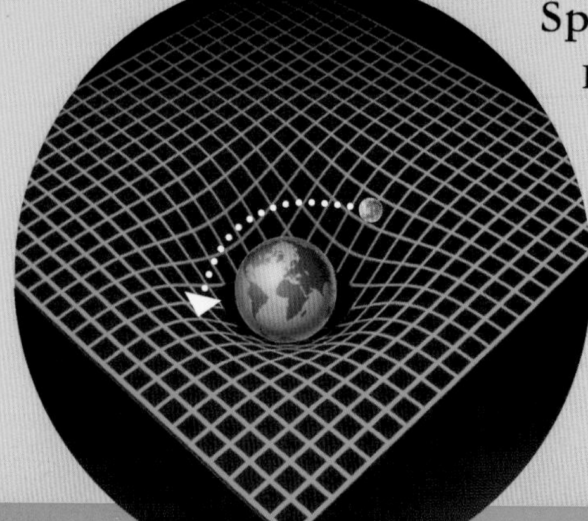

Space bends

Einstein's theory is called the theory of relativity because how things look depends on how you're moving relative to them. In the world of relativity, gravity is different too. We usually think of gravity as a force that tugs things to the Earth, but in Einstein's theory gravity happens for another reason. Heavy objects – like planet Earth – *bend space and time*. This is easier to understand if you imagine space and time as a sheet of stretchy rubber. If you put planet Earth in the middle, it makes a dip. Now roll the Moon past. It follows the curve of space and gets trapped in orbit around Earth because it can't climb out of the dip.

What you *see* depends on how

at the speed of light?

90%

... Now you're doing 90% of light speed and things are getting weird. To the people outside, your car has shortened to half its length and time has slowed down inside your car. From your point of view, the spectators have become narrower and slower, and your car seems fine.

99%

At 99 per cent of light speed, your car is less than a metre long, you're seven times heavier, and a day for you lasts a week to the people outside. They now look as thin as rakes and are moving in slow motion. When they speak, their words come out increddddddibly slowwwwwwly.

99.99%

At 99.99% of light speed, your car is shorter than a pencil and a second lasts more than a minute. Going faster just makes you shorter and slows time even more. You can never reach the speed of light – it's impossible. If you could, you'd be zero length, heavier than the Universe, and time would stop.

you'd be heavier than the UNIVERSE

Gravity changes time

Einstein discovered that gravity slows down time. At the top of a mountain, where Earth's gravity is very slightly weaker, time runs a little quicker. But only a *tiny bit* quicker. Even if you spent your entire life at the summit of Mount Everest, you'd get older by only a tiny fraction of a second more than your friends at sea level.

Time travel

According to Einstein, travelling forwards in time is easy. Just jump in a rocket, whizz round the Universe at 99.99% of the speed of light (or hang around something with colossal gravity, like a black hole), and come back in 4 months. Everyone you know will be 24 years older!

you're moving – that's RELATIVITY

Who's who?

The brilliant scientist and mathematician Sir Isaac Newton once said, "If I have seen farther, it was by standing on the shoulders of giants." Newton meant that his own work was built on the work of the great scientists who lived before him. Here are some of the biggest names in physics, starting back in ancient Greece.

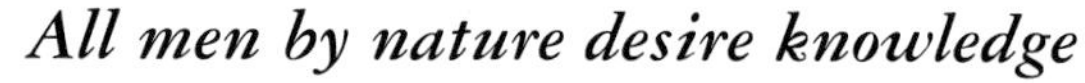

ARISTOTLE 384–322 BCE

One of the great thinkers of the ancient world, the Greek philosopher Aristotle was an expert on subjects from anatomy and astronomy to physics and philosophy. He was a tutor to Alexander the Great and a lecturer at the greatest school in Greece. He valued knowledge gained from observing nature and said, "Nature does nothing uselessly".

ARCHIMEDES 287–212 BCE

A Greek mathematician, astronomer, physicist, and engineer, Archimedes invented war machines and a device to lift water uphill. His most famous moment (when he sprang out of the bath shouting "Eureka!") came when he discovered that the king's crown wasn't solid gold, because it displaced more water than the same weight of pure gold.

COPERNICUS 1473–1543

The Polish astronomer Nicolas Copernicus believed that the Sun, not the Earth, is the centre of the Universe, and that the planets revolve around the Sun. Although this idea offended the ruling Catholic Church, his work helped later scientists unravel the forces that govern the Universe, and it formed the basis of modern astronomy.

GILBERT 1544–1603

A distinguished London doctor (and physician to Elizabeth I), William Gilbert is best known for his research into electricity and magnetism. In his book *De Magnete* (*On Magnetism*), he explained how magnets attract and repel. He also showed that Earth is like a giant bar magnet, which is why a compass needle always points north.

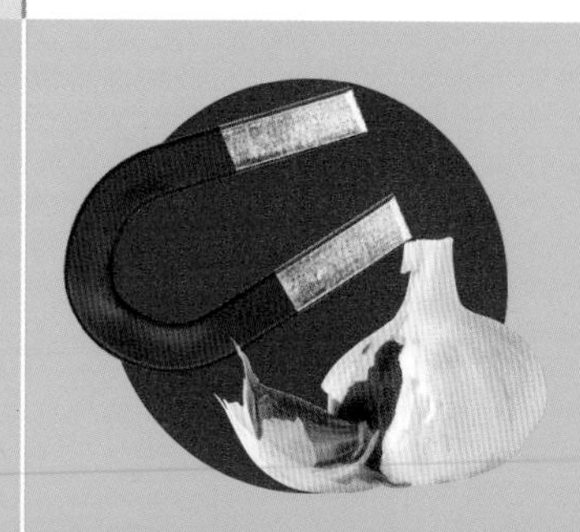

GALILEO 1564–1642

The Italian scientist Galileo Galilei used a telescope of his own design to study the Moon, Sun, stars, and to discover Jupiter's moons. He rolled balls down ramps in experiments on gravity and discovered that all falling objects accelerate at the same rate. His views on astronomy led the Church to place him under house arrest for the last years of his life.

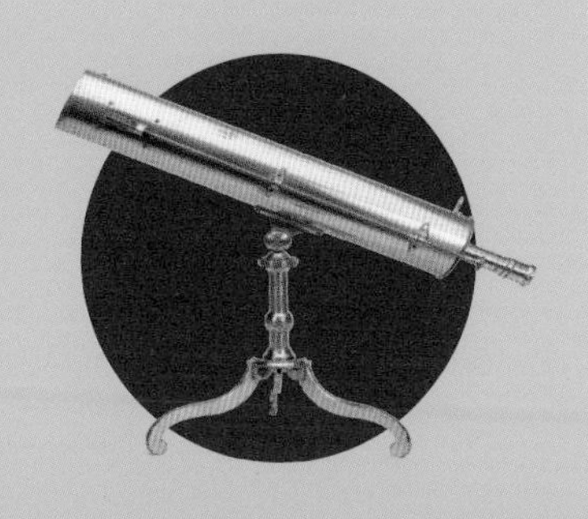

NEWTON 1642–1727

The English scientist Isaac Newton often tops polls as the greatest scientist ever, but he had strange religious ideas and believed in alchemy (a magical form of chemistry). Best known for working out how gravity holds the Universe together, he also published theories of colour. He was very sensitive to criticism and fell out with many of his contemporaries.

FRANKLIN 1706–1790

One of the founding fathers of America, Benjamin Franklin was a prominent politician and diplomat as well as a scientist. He discovered that lightning is a form of electricity by flying a kite in a storm, and he invented the lightning conductor. He was the first person to describe electricity as having positive and negative forms.

VOLTA 1745–1827

Italian physicist Alessandro Volta did not speak until he was four but soon caught up with his classmates. He invented the first battery by discovering that an electric current could flow between two metals in a solution. His fame spread and he performed experiments in front of Napoleon. The basic unit of electric force was named the volt in his honour.

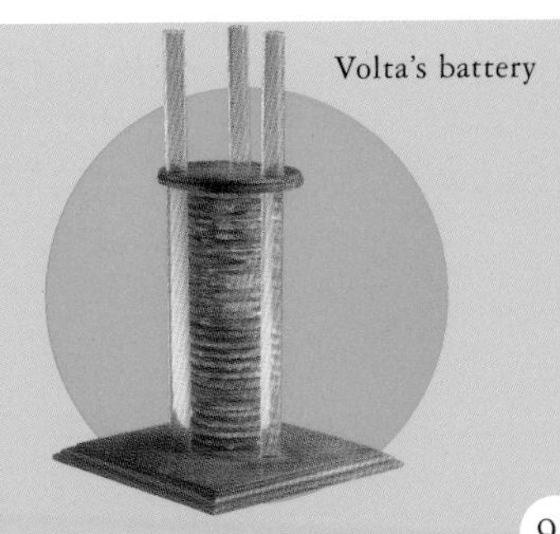
Volta's battery

Nothing is too wonderful to be true if it be consistent with the laws of nature

One never notices what has been done; one can only see what remains to be done

If the facts don't fit the theory, change the facts

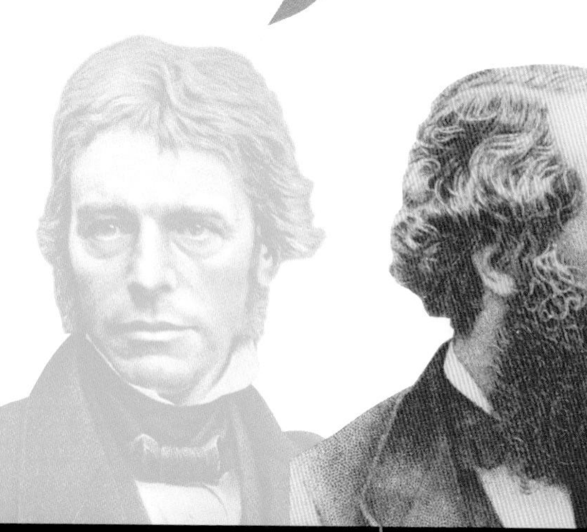

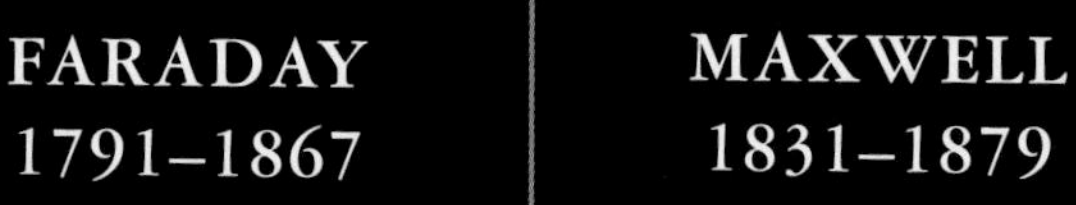

FARADAY 1791–1867

The English physicist Michael Faraday was a brilliant experimenter. He discovered that electricity and magnetism are closely linked, and he built the first electrical generator. An enthusiastic public speaker, he founded the tradition of Christmas science lectures that continues at England's Royal Institution to this day.

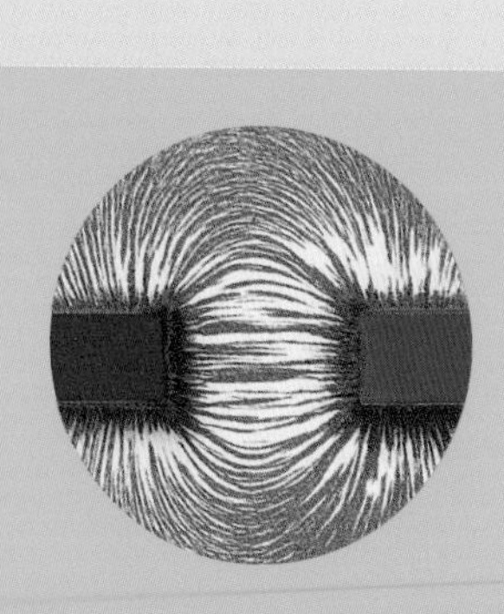

MAXWELL 1831–1879

Nicknamed Dafty as a boy, James Clerk Maxwell of Scotland grew up to be a brilliant physicist, famous for summing up electromagnetism in four short equations. He showed that electricity and magnetism can travel as waves and he proved that light is a form of electromagnetic radiation. He also showed that heating a gas makes the atoms move faster.

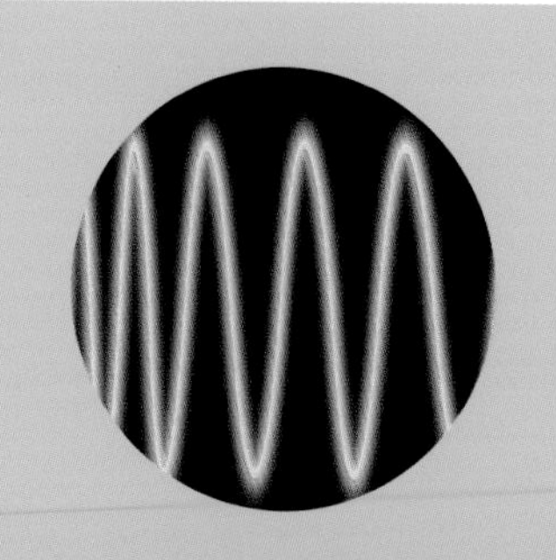

CURIE 1867–1934

The first female Nobel Prize winner, Polish-born Marie Curie is famous for her work on radioactivity. She discovered two radioactive elements and worked out how to purify one of them (radium). The discoveries led to a new treatment for cancer: radiotherapy. Ironically, Marie Curie died of a type of cancer caused by exposure to radioactivity.

EINSTEIN 1879–1955

Albert Einstein's first success was to show that light is made of particles (photons), but he is better known for his theory of relativity, which links space, time, and gravity with the speed of light. His formula $E = mc^2$ shows that energy and mass are the same thing. Einstein opposed war, but his work led to the invention of nuclear bombs, which convert mass into energy.

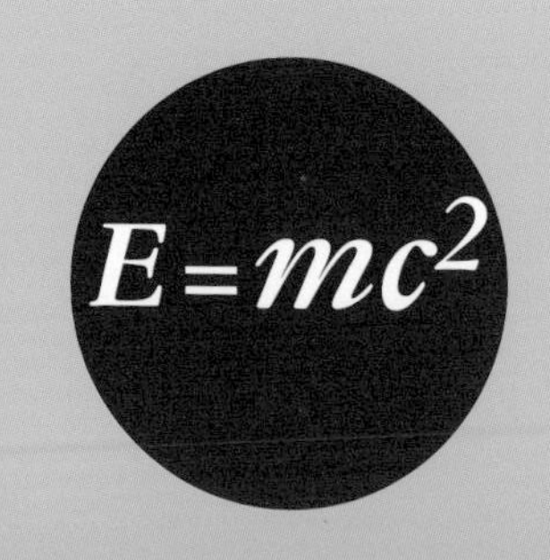

All science is either physics or stamp collecting

What we observe as material bodies and forces are nothing but shapes and variations in the structure of space

In science one tries to tell people, in such a way as to be understood by everyone, something that no one ever knew before

RUTHERFORD 1871–1937

Ernest Rutherford was a fiery New Zealander who "split the atom". While working in England he proved that atoms are made from smaller parts, and he discovered that they contain a tiny solid centre, which he called the nucleus. He was one of the first people to investigate radioactivity – work that laid the foundations of atomic physics.

SCHRÖDINGER 1887–1961

Austrian physicist Erwin Schrödinger worked out the complex maths needed to show how atoms and the particles within them can behave as waves. He is best known for a famous puzzle in which the bizarre quantum world inside an atom is magnified to everyday dimensions, causing a cat to be both dead and alive at the same time.

HEISENBERG 1901–1976

German physicist Werner Heisenberg helped to figure out the puzzling inner world of the atom. At 23 he developed his theory of quantum mechanics, which won him a Nobel Prize. He also discovered the strange "uncertainty principle", which says it's impossible to know both the location and speed of a subatomic particle at the same time.

DIRAC 1902–1984

The English scientist Paul Dirac helped develop theories to show that subatomic particles like the electron behave as waves. Dirac's work helped to advance quantum mechanics. He also showed that a negatively charged electron must have a positively charged twin, a positron. He shared the Nobel Prize for physics with Schrödinger in 1933.

GLOSSARY

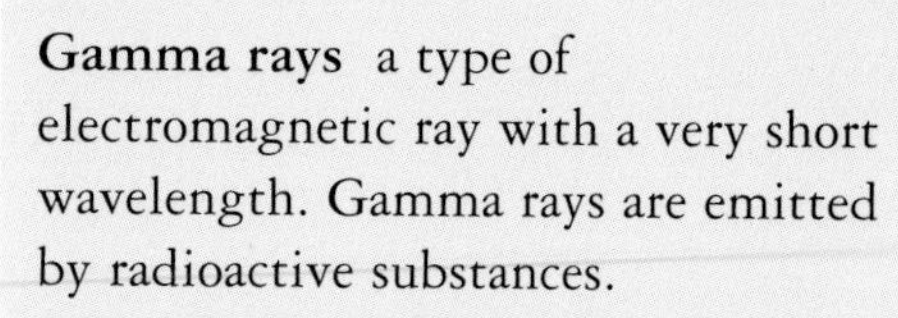

Absolute zero the lowest temperature possible (–273°C or –459°F), when all atoms stop moving.

Acceleration how quickly an object speeds up, slows down, or changes direction.

Aerodynamics the study of how objects move through air.

Atom a tiny particle of matter, consisting of a central nucleus surrounded by one or more electrons.

Battery a device that can produce or store electricity.

Buoyancy a force that pushes upwards on an object when it's surrounded by liquid or gas.

Celsius a temperature scale based on the freezing point (0°C) and boiling point (100°C) of water.

Centre of gravity a point in an object where all its weight appears to be concentrated. You can balance a pencil halfway along its length because that's where its centre of gravity is.

Charge (or electric charge) the amount of unbalanced electricity in an object. Charge can be negative (too many electrons) or positive (not enough electrons).

Circuit the path through which an electric current can flow.

Conduction the movement of heat, sound, or electricity through a substance.

Conductor a substance that transmits heat, sound, or electricity well.

Convection the transfer of heat through a fluid (liquid or gas) by moving currents.

Drag the force of resistance an object experiences when it travels through a gas or liquid.

Elastic an object is elastic if it returns to its original size and shape after being stretched.

Electric current the flow of electric charge (usually in the form of electrons) through a circuit.

Electricity effects caused by the movement of electric charges.

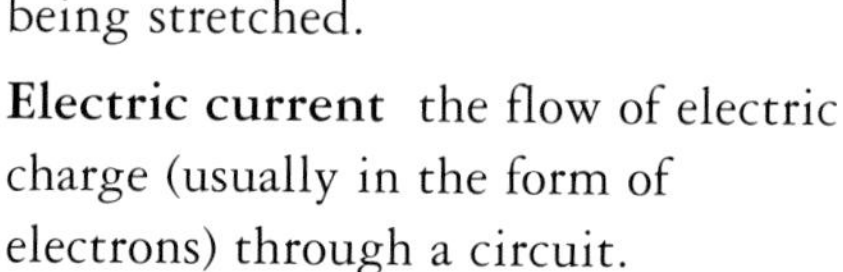

Electromagnetic ray a form of energy that travels at very high speed and can act as either waves or particles. Light and radio waves are types of electromagnetic ray.

Electron one of the three main particles in an atom (with the proton and neutron). It has a negative charge.

Energy the ability to do work (*see* work).

Evaporation a change of state in which a liquid turns into a gas (vapour).

Fluid a substance that flows. Liquids and gases are fluids.

Force a push or a pull. Forces change the speed, direction, or shape of objects.

Freezing point the temperature at which a liquid changes into a solid.

Frequency the number of waves that pass by each second.

Friction a force that hinders motion when one surface moves over another.

Gamma rays a type of electromagnetic ray with a very short wavelength. Gamma rays are emitted by radioactive substances.

Gas a state of matter in which the particles are far apart and move about randomly and quickly.

Gravity a force of attraction between all objects. Earth's gravity keeps our feet on the ground and makes objects fall when we drop them.

Heat a form of energy caused by the random motion of atoms.

Inertia the tendency of an object to resist any change in its motion.

Infrared ray a type of electromagnetic ray with a longer wavelength than visible light. Infrared rays transmit heat.

Insulator a material that is poor at conducting heat, sound, or electricity.

Interference a disturbance produced when two or more waves (such as light or sound) collide.

Lift the upward force produced on a wing when it moves through the air.

Light a type of electromagnetic ray that our eyes can see. White light is a mixture of all the colours of the rainbow, which together make up the visible spectrum.

Liquid a state of matter between a solid and a gas, in which the particles can slide around but remain close together and attract one another.

Magnetic field the region around a magnet, where its forces act.

Magnetism the property of some materials, especially iron, to attract or repel similar materials.

Mass the amount of matter in an object, measured in grams, kilograms, or tonnes. On Earth, an object's mass gives it weight. In space, objects are weightless but still have mass.

Matter anything that has mass and occupies space.

Melting point the temperature at which a solid turns into a liquid.

Microwaves electromagnetic rays with a wavelength longer than that of infrared rays but shorter than that of most radio waves. Microwaves are sometimes said to be a type of radio wave.

Molecule a particle of matter made of two or more atoms strongly bonded together.

Momentum the tendency of an object to keep on moving, equal to its mass times its velocity.

Neutron one of the two main particles in the nucleus of an atom. It has no electric charge.

Nucleus the solid centre of an atom, made up of protons and neutrons. It contains most of the atom's mass.

Particles basic units from which all substances are made, such as atoms or molecules. Subatomic particles are those smaller than an atom, such as protons.

Photon a particle of light or of any other type of electromagnetic ray.

Physics the scientific study of force, motion, matter, and energy.

Potential energy energy stored for later use. Gravitational potential energy is the energy stored in an object because of its position, such as a rollercoaster at the top of a hill.

Pressure how concentrated a force is over an area.

Prism a triangular wedge of glass or other transparent material that can split white light into a spectrum of colours.

Proton one of the two main particles found in the nucleus of atoms. It has a positive charge.

Quantum a single, tiny unit of energy. Electromagnetic rays consist of a stream of quanta called photons.

Quark one of the ultimate building blocks of all matter. Protons and neutrons are made of three quarks each.

Radiation energy travelling as electromagnetic rays (such as light). The various types of ray given off by radioactive substances are also called radiation.

Radioactivity high-speed particles or electromagnetic rays given off when atomic nuclei break down.

Refraction the bending of a beam of light as it passes from one substance to another, such as from air to water.

Sound a kind of wave that travels through air or other materials, caused by molecules being briefly squeezed together.

Spectrum a succession of electromagnetic rays arranged in order of their wavelength, from shortest to longest. A rainbow is a spectrum of visible light wavelengths.

Speed how fast an object is moving, calculated as distance divided by time.

Streamlining shaping an object so that it travels with the least resistance, or drag, through air or water.

String in string theory, strings are the ultimate building blocks of the Universe. Particles are vibrations on the string.

Strong nuclear force the force that holds the nucleus of an atom together. It is felt only by protons and neutrons and, unlike gravity or magnetism, acts only over very short distances.

Subatomic particle a particle that is smaller than an atom, such as a proton, electron, or quark.

Surface tension a force that causes the surface of water to form a weak skin.

Temperature a measure of how hot or cold something is.

Ultraviolet (UV) light invisible electromagnetic ray with a shorter wavelength than visible light. UV rays in sunlight can burn skin.

Velocity the speed and direction of an object.

Viscosity A measure of how thick or runny a liquid is.

Wavelength the distance between the crest of one wave and the next.

Weight the downward force of an object's mass due to Earth's gravity.

Work a measure of the amount of energy used when something happens, equal to force times distance.

X-rays Electromagnetic rays with wavelengths between those of gamma rays and ultraviolet rays. X-rays pass through most parts of the human body, but not bones or teeth.

Zero-G alternative term for weightlessness.

INDEX

absolute zero 65
acceleration 24-25, 26, 31, 40, 47
aerodynamics 46-47
aerofoil 42
air 51, 84, 85
Alhazen 13
amber 8
apple 18-19
Archimedes 10, 92
Aristotle 9, 92
atmosphere 84, 85
atom 8, 11, 19, 54-55, 56-57, 58-59, 62, 64, 66, 67, 76, 77, 82, 83
Bernoulli effect 43
bicycle 28, 29, 37, 38-39
bubble 80-81
bullet 44-45
cannonball 16, 19
car 21, 29, 37, 46-47
clouds 85
colour 78-79, 80-81, 82, 84, 85
compass 12, 63
condensation 67
conduction 64
convection 64
Copernicus, Nicolas 14, 15, 92
distance 37
drag 26, 39, 40, 41, 43
Earth 10, 11, 14, 15, 17, 18, 19, 31, 62, 84
efficiency 38
Einstein, Albert 35, 87, 88, 89, 94
elastic limit 73
elasticity 48-49
electricity 58-59, 60-61 63
electromagnetic spectrum 82-83
electromagnetism 82-83
electron 57, 58, 59, 60, 61, 62, 63
energy 34-35, 48, 49, 66-67
evaporation 67
feather 8, 9
fluid 70-71
football 45, 49
force 9, 20-51
force field 62, 77
freezing 66
friction 9, 25, 26, 27, 28-29, 30, 39, 40, 70
g force 31, 32-33
Galileo 16-17, 93
gamma rays 82
gas 51, 66, 67
gears 37
Gilbert, William 15, 92
gold 10, 19
golf ball 44, 49
gravity 9, 16, 18, 19, 23, 24, 26, 27, 31, 33, 40, 88, 89
Grimaldi 77
heat 83
Hero 11
inertia 16, 17, 19, 25, 26, 31, 38, 45
infrared (IR) wave 83
iron 8
laws of motion 19, 24-25, 26-27, 43, 72
lever 10, 36-37, 39
lift 42-43
light 13, 74-89
light waves 76, 77, 78, 80, 82
lightning 59
liquid 51, 66, 67, 80
lodestone 8
machines 36-37
magnet 12, 13, 15, 62, 63
magnetism 8, 62-63
Magnus effect 45
maths 19
melting 66
microwave 83
Moon 11, 14, 17, 18, 19, 84
motion 16
needle 12, 15
neutron 57
Newton, Isaac 18-19, 24, 25, 70, 76, 79, 93
nucleus 57, 58
orbit 19
parachute 41
particle 76, 77
Peregrinus 13
photon 76
planets 11, 14, 24
plasma 66, 67
pressure 50-51
prism 78, 79
proton 57
quark 57
radiation 64
radio wave 83
radioactivity 82
rainbow 78
refraction 87
religion 12, 14, 15
scattering 84
sea 85
solid 66, 67
sound waves 76
spectrum 78, 79
speed 30-31
speed of light 86-87, 88-89
stars 10, 11, 17, 67, 85
static electricity 58, 59
Sun 10, 11, 13, 14, 15, 17, 18, 19, 31, 84, 85
sunlight 78, 82, 84
surface tension 80
terminal velocity 40, 41
theory of relativity 88
time 88, 89
traction 28
turbulence 44, 46, 47
TV 78, 83
ultraviolet (UV) light 82, 84
Universe 10, 11, 14-15, 18-19
velocity 30-31
viscosity 70-71
vortex 43
water 51, 68-69
wavelength 80, 82-83, 84, 85, 78
X-ray 82
Young, Thomas 77

Acknowledgements

Dorling Kindersley would like to thank the following people for help with this book: Tory Gordon-Harris, Rose Horridge, Anthony Limerick, Carrie Love, Lorrie Mack, Lisa Magloff, Rob Nunn, Laura Roberts-Jensen, Penny Smith, Fleur Star, Sarah Stewart-Richardson, Likengkeng Thokoa.

DK would also like to thank the following for permission to reproduce their images (key: a=above, b=below, c=centre, l=left, r=right, t=top, b/g=background):

Advantage CFD: 46-47b. **Alamy Images:** Bruce McGowan 66bc; Ian M Butterfield 37bl; Pat Behnke 12tl (Diamond); Robert Llewellyn 73bl; Doug Steley 66bl; TF1 33crb, 46bl; Transtock Inc. 46tl, 47crb; Stephen Vowles 41 (soldier), 94 (soldier). **Ariel Motor Company:** 47tr, 95br. **Corbis:** 30bl; Car Culture 31c; 33cra, 49 (golf ball), 71tc; Theo Allofs/zefa 67cla; Bettmann 18crb, 24bl, 27bl, 40-41b, 42tc, 59bl; Tom Brakefield 31 (cheetah); Ralph A. Clevenger 69tr; Richard Cummins 39cr, 66-67b; Ronnen Eshel 2bl, 26bl; Eurofighter/Epa 42tl; Al Francekevich 39br; Patrik Giardino 22cr; Beat Glanzmann/zefa 78cl; Images.com 5cla, 20-21; JLP/Jose Luis Pelaez/zefa 3 (hand), 23tr (hand); Kurt Kormann/zefa 79tr; Matthias Kulka/zefa 34tl; Lester Lefkowitz 42-43; George D. Lepp 47br; LWA-Sharie Kennedy 29 (moped); David Madison/zefa 40cl; Tim McGuire 38c; Amos Nachoum 51cra; Yuriko Nakao/Reuters 49br; Richard T. Nowitz 23cl; Robert Recker/zefa 83tc; Reuters 29cb, 30cl, 35cr, 47tr, 47cra; Galen Rowell 64tr; Wolfram Schroll/zefa 39c; G. Schuster/zefa 48br; Daniel Smith/zefa 66tc; Josh Westrich/zefa 3 (trolley), 25tl. **Deutsches Museum, München:** 31 (helios). **DK Images:** Adidas 1 (trainer); The All England Lawn Tennis Club, Church Road, Wimbledon, London, David Handley 48 (tennis ball); Anglo-Australian Observatory 14t (b/g), 17tc (b/g), 22tl (b/g); Audiotel 1 (spanner); Bradbury Science Museum, Los Alamos 34bl, 55bc; British Airways 1 (Concorde); The British Museum 60b (silk); Duracell Ltd 58cla; Ermine Street Guard 8bc; Football Museum, Preston 45 (boots), 61b (shirt); Sean Hunter 2fcl, 16tl; Indianapolis Motor Speedway Foundation Inc 36cb; NASA 1 (astronaut), 10tr, 18bl; National Maritime Museum, London 16cr, 91fbl; Natural History Museum, London 9tc, 19fcrb, 90 (feather); Stephen Oliver 1 (domino), 13cra, 25 (car), 36c, 48 (basketball), 49 (basketball), 52bl, 62-63, 63bc, 63bl, 92fbl; Renault 88-89t; The Science Museum, London 11bl, 12fbr, 19bl, 50-51 (nails), 58bl, 91fbr; James Stevenson/National Maritime Museum, London 15cr; Florida Center for Instructional Technology: 9br. **Dreamstime.com:** Cosmin-Constantin Sava 83cb (Modern mobile phone), Jamie Cross 83crb, Sacanrail 83cb. **Getty Images:** Allsport Concepts 22cl; Iconica 68cl; The Image Bank 23cr, 23bl, 74bl; Nordic Photos 31 (Caravan); Photographer's Choice 36tl, 48tl; Stock Illustration Source 87tl; Stone 38br; Stone+ 58fcla; Taxi 1 (skydiver), 2tl, 40-41t, 74-75 (Sun), 85cla. **Carrie Love:** 8cra (b/g). **Mary Evans Picture Library:** 28bc. **NASA:** Marshall Space Flight Center 1 (Shuttle). **Photolibrary:** Photolibrary.Com (Australia) 17br; Foodpix 35tr, 53tl, 54bl, 70tc; Index Stock Imagery 5tl (b/g), 6-7 (b/g), 50bc. **PunchStock:** Corbis 17cr. **Photo Scala, Florence:** 16cra. **Joe Schwartz/Joyrides:** 32r. **Science & Society Picture Library:** Science Museum Pictorial 29bl. **Science Photo Library:** 10br, 10cla, 54br, 90fcla, 90cla, 90bl, 92fcla, 92cla; American Institute Of Physics 93fcra; David Becker 39bl; George Bernard 91cra; British Antarctic Survey 82crb; Dr. Jeremy Burgess 38tl, 80-81b; CERN 76br; John Chumack 85bl; Crawford Library/Royal Observatory, Edinburgh 15bc; Professor Harold Edgerton 45bc, 72-73t; Prof. Peter Fowler 93fcla; Mark Garlick 67br; Henry Groskinsky, Peter Arnold Inc. 48c; GUSTO 74tc; Roger Harris 15tr, 62tl; Keith Kent 31 (Thrust); Edward Kinsman 72c; Mehau Kulyk 5bl, 74tl; Laguna Design 88bl; Damien Lovegrove 74bc; Max-Planck-Institut/American Institute Of Physics 93cra; NASA 31 (Apollo15), 33bl; National Library Of Medicine 92cra; Claude Nuridsany & Marie Perennou 74br; David Parker 74-75, 79cla, 81cr (bubble); Pasieka 74-75b, 77tl; D. Phillips 52tc, 65tl; Philippe Plailly 52-53b, 54bc; D. Roberts 82clb; Royal Observatory, Edinburgh 18-19 (map); Erich Schrempp 80cr; Science, Industry & Business Library/New York Public Library 17tr; Dr. Gary Settles 44bl; Francis Simon/American Institute Of Physics 93cla; Sinclair Stammers 80tr, 81cl, 81cb; Takeshi Takahara 46cb; Ted Kinsman 83fclb; Sheila Terry 3 (Atlas), 14bl, 63fcra, 70clb, 79cb, 91fcla, 91cla, 91fcra; Gianni Tortoli 3 (heliocentric), 15cl; US Library Of Congress 35br, 86br, 92fcra; Detlev Van Ravenswaay 19cr, 90cra, 90br. **Sky TV:** from Sky One's Brainiac 70-71b.

For further information see: www.dkimages.com